**MANWEB**
**CUSTOMER INTELLIG**
(POWER MARK

Room 2S1, Head Office, Chester.

# Questions of power

*For Yvonne and in memory of Daniel*

# Questions of power

## Electricity and environment in inter-war Britain

Bill Luckin

Manchester University Press
Manchester and New York
Distributed exclusively in the USA and Canada by St. Martin's Press

Copyright © Bill Luckin 1990

*Published by* Manchester University Press
Oxford Road, Manchester M13 9PL, UK
*and* Room 400, 175 Fifth Avenue,
New York, NY 10010, USA

*Distributed exclusively in the USA and Canada
by* St. Martin's Press, Inc.,
175 Fifth Avenue, New York, NY 10010, USA

*British Library cataloguing in publication data*
Luckin, Bill
    Questions of power: electricity and environment in inter-war Britain.
    1. Great Britain. Electricity supply, history
    I. Title
    621.310941

*Library of Congress cataloging in publication data*
Luckin, Bill
    Questions of power: electricity and environment in inter-war Britain/Bill Luckin.
    p. cm.
    Includes bibliographical references.
    ISBN 0-7190-3302-0
    1. Electric utilities—Great Britain—History—20th century.
2. Electric power-plants—Environmental aspects—Great Britain—History—20th century. 3. Great Britain—Economic conditions—1918–1945. I. Title
HD9685.G7L83  1990
333.79'32'094109041—dc20                                90-31179

ISBN 0 7190 3302 0 *hardback*

Phototypeset in Monotype Garamond
by Megaron, Cardiff, Wales

Printed in Great Britain
by Biddles Ltd, Guildford and King's Lynn

# Contents

|   | List of illustrations | vi |
|---|---|---|
|   | Preface | vii |
|   | Introduction | 1 |
| **Part I** | **Adoption** | |
| 1 | The meanings of triumphalism | 9 |
| 2 | Constructing the image | 23 |
| 3 | Targeting women | 39 |
| 4 | Urban experiences | 52 |
| 5 | Rural stagnation | 73 |
| **Part II** | **Resistance** | |
| 6 | Downs, lakes and forests | 94 |
| 7 | Exploiting the Highlands | 118 |
| 8 | The Battersea controversy | 138 |
| 9 | Arcadia under threat | 156 |
| 10 | Nuclear aftermath | 172 |
|   | Note on sources | 186 |
|   | Select bibliography | 190 |
|   | Index | 196 |

# Illustrations

|   |   | page |
|---|---|---|
| 1 | The structure of the electrical industry in inter-war Britain | 2 |
| 2 | Domestic electricity supply in Britain, 1924–40 | 10 |
| 3 | Manchester in the inter-war years | 55 |
| 4 | Growth of electricity sales in Manchester | 56 |
| 5 | The National Grid as planned, c. 1930 | 93 |
| 6 | Centres of electrical debate in the South Downs and adjacent areas, summer to winter, 1929 | 97 |
| 7 | The area at issue during the Keswick debate, 1929–33 | 103 |
| 8 | Foci of electrical debate and literature in the New Forest region, 1930–34 | 110 |
| 9 | The area of Inverness-shire most directly affected by the Caledonian schemes | 120 |
| 10 | Institutions and areas 'under threat' during the Battersea crisis | 147 |

# Preface

This study, which is primarily concerned with electricity in the domestic sphere, was originally planned as a co-authored project with Russell Moseley, of the Council for National Academic Awards, and now of the University of Warwick. When pressure of work forced Dr Moseley to withdraw I continued to benefit from his advice and support. He played a decisive role in helping me to shape the study as a whole, read and commented on every chapter, and allowed me to make use of his research notes for the section on Battersea Power Station. I am very grateful to him. I must also thank the Economic and Social Research Council for inviting Dr Moseley and myself to participate during 1983–4 in a multi-disciplinary investigation into 'Public Attitudes towards New Technologies': it was during that period that I first identified the problematics that lie at the heart of this book, and *via* a chance viewing of an American made-for-television movie – *O.H.M.S.* (1976, directed by Dick Lowry) – became convinced that there must have been a British anti-pylon movement. I would like to thank those members of the Public Attitudes group who commented on working papers by Dr Moseley and myself at that early stage.

I have also been generously supported by the University of Manchester. During 1987–8 I was appointed Simon Senior Research Fellow in the Centre for the History of Science, Technology and Medicine, and this provided an excellent and congenial base for writing and research. John Pickstone has been a generous academic and personal host and reader of draft chapters. Joan Mottram has helped with a number of essential practical matters. David Edgerton has shared his wide knowledge of British inter-war history. At an early stage, Jon Harwood made astute comments on a draft chapter. I also want to thank Roger Cooter. He has listened, probably more often than he cares to remember, to my ruminations on content, style, structure and history in general – his advice has always been constructive and down-to-earth. John Walton has sent me references, imparted his deep knowledge of urban–rural relations in the North-West, and provoked me into asking questions that I would have otherwise ignored. Malcolm Pittock, at Bolton, read early chapters with his usual assiduity: he lived through part of the 'electrical revolution' and, as a result, his comments were refreshingly concrete. The same goes for Jack Beeching, who took valuable time away from novel writing to send me a long letter about electrical and political consciousness in Sussex in the 1930s. Ian Welsh kindly let me see a chapter from his thesis – due acknowledgement is made in the notes at the end of the book. Peter Searby gave encouragement and reassurance when it was most urgently needed and the same is true of Chris Gibson, Frank Margison, Patricia Wood and Penny Eyles who provided an ever-welcoming *pied-à-terre* in London. Thanks are due also to Michael Fitch for his meticulous copy-editing and to Isabel Hayley, at Manchester University Press, for expertly overseeing the production process. Finally, the research for the rural sections of the book involved reading about and entering into a world that had long since been brought vividly alive for me by my late father, Geoffrey Luckin, who lived and farmed in Essex during the inter-war years. In one way or another his humane testimony has influenced everything that I have written here about farming and the countryside.

Among librarians and archivists, the unsung heroes of every form of historical scholarship, I must single out Bridget Malcolm for her unstinting help at the Electricity Council; officers and staff at the Central Electricity Generating Board Archive, Bankside House, Southwark; the Museum of English Rural Life at the University of Reading; the Cumbria and Hampshire County Record Offices; the Council for the Protection of Rural England; Nuffield College, Oxford; the Institution of Electrical Engineers; the Local History Library at the Central Reference Library, Manchester; Leeds City Library; the British Library at Bloomsbury, Colindale and Boston Spa (where I was particularly grateful to be allowed to wander among the pre-war electrical shelves); and the Public Record Office at Kew.

# Introduction

For more than a century and a half ambivalence and opposition to technological change have lain at the heart of British social life. Deploring the new modes of production which precipitated the final demise of the domestic system, and decrying the 'harsh, inhuman' lines of the railway, nineteenth-century critics of mechanisation depicted such innovations as disruptive of both the natural and the social orders. For their part, apologists for the onward rush of technical change, and its motor force, unfettered bourgeois capitalism, dismissed such attitudes as backward-looking and 'feudal' and insisted that what are now called the social benefits of large-scale technological and productive systems would always outweigh the short-term costs.

Those who sought to protect the traditional patrician order were castigated as unenlightened conservatives opposed to a more equal distribution of goods and jealously protective of rural property rights. On the other side of the cultural divide, advocates of 'mere mechanical progress' were dubbed materialists and philistines. This complex and seemingly unending argument was and still is locked into two other historical processes. The first is the debate about the respective merits, and relative moral values, of town and country. The second is the struggle, stretching from the communal defence of the 'just price' through the Luddites and the travails of the handloom weavers, to recent attempts to combat the unilateral imposition of cost-cutting 'rationalisations' on organised labour. Here, and this is a theme which will recur, the social history of technology impinges upon and interpenetrates with political and economic history.

The conflict between traditionalists and 'triumphalists' which was triggered by the construction of the National Grid between 1927 and 1934, and which provides the central focus for this book, lay in lineal

```
                    Ministry of Transport


 Central Electricity Board ──────── Electricity Commissioners
      (National Grid)                  (Technical and legal)


  Private and municipal
    supply companies


                              Publicity and 'propaganda'
                               (i) Electrical Development Association
                               (ii) Electrical Association of Women
                               (iii) Electrical press
```

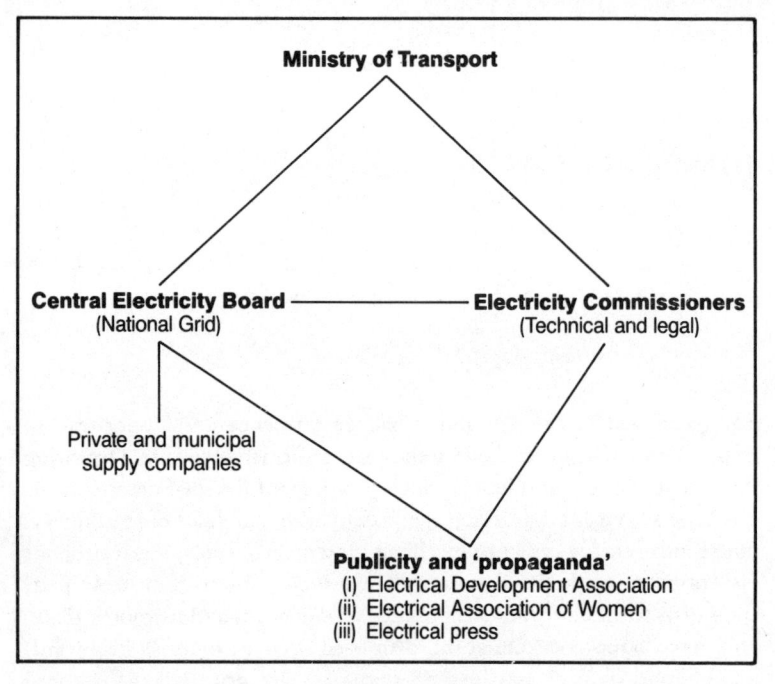

**Figure 1** The structure of the electrical industry in inter-war Britain

descent from these nineteenth-century mechanisation debates. Triumphalism was grounded in the scientific premiss that economy, society and culture would be rapidly and radically transformed by the new source of energy. Forged by electrical engineers, contractors, salesmen, journalists and technocrats, this progressivist ideology was buttressed by powerful state bureaucracies. It relied on the support of the linked triad of the Ministry of Transport (which was responsible for electrical legislation), the Electricity Commissioners and the Central Electricity Board (CEB). The Ministry maintained a low parliamentary profile throughout the inter-war period, preferring, under every administration except Labour between 1929 and 1931, to play an enabling rather than an interventionist role. The Electricity Commissioners were modelled on the Railway Commissioners and had been responsible since 1919 for the legal and technical aspects of new schemes: from 1927 they co-operated closely with the Central Electricity Board, a formally independent body, whose statutory role was to construct and administer the Grid.

The progressivist tenets of triumphalism were also advanced by a public relations organisation, the Electrical Development Association (EDA), which had grown out of a subcommittee of the Institution of Electrical Engineers in the immediate aftermath of World War I. The EDA attempted to disseminate positive images of the new energy source and developed close links both with the CEB and with another quintessentially triumphalist organisation, the Electrical Association for Women (EAW). The brain-child of an electrical engineer, writer and propagandist, Caroline Haslett, the EAW had been founded in 1925 as an offshoot of the Women's Engineering Society; it was dedicated to the demystification of electricity for every housewife and 'bachelor girl' in the land.

Each of these bodies – the Ministry of Transport, the Central Electricity Board, the Electricity Commissioners, the Electrical Development Association and the Electrical Association for Women – played decisive roles in defining and refining triumphalism. They were assisted by an outspoken technical press: the *Electrician, Electrical Times, Electrical Review*, and, most scathing of all in its indictment of 'reactionaries' and 'anti-electrical elements', *Electrical Industries and Investments*. Since they are central to an understanding of the ideological texture and cultural resonance of electrical progressivism, the EDA and EAW are given detailed consideration in the first part of this book. In the chapters which follow, on electricity in town and countryside in Britain during the 1920s and 1930s, the limits and contradictions of triumphalist orthodoxy are defined and the idea of electrical utopianism juxtaposed against the grass-roots realities of domestic supply. Since little is known about precisely which localities, social classes, and types of houses had full or partial access to the new energy source, the chapter on 'Urban Experiences' is devoted to an in-depth study of a single city, Manchester. It concentrates, in particular, on the factors which determined the spatial and social dissemination of the new source of power and the continuing commercial and cultural attractiveness of gas.

If, in terms of firmly established facts and figures, the social history of urban supply remains underdeveloped, the spread of electricity in the countryside is very nearly *terra incognita*. An attempt is nevertheless made in Chapter 5 to provide an estimate of the extent of rural provision. Progressives were convinced that the new energy source would play a crucial role in bringing an end to agricultural depression. This 'techno-arcadian' campaign for the mechanisation of rural crafts and skills and the revivification of village communities allegedly bled white by long-term depopulation, represents yet another aspect of the Janus-like face of triumphalism. Judged by results, this movement was no more successful than the earlier passion for 'electro-culture' which sought to cultivate arable fields with electric ploughs and deploy ultra-violet to induce dramatic improvements in the growth rates of orchids, lilies and tomatoes.

The future of mass electricity in Britain was deeply influenced by events in the countryside between 1929 and 1934. The battles that were fought in the full glare of national publicity in the South Downs, the Lake District, the New Forest, and the Scottish Highlands, pitted localities and regions against central bureaucracies. Rural preservationism and 'amenity' were crucial issues but so, also, were anti-centralism and the right to be heard before unprejudiced public inquiries. That utilitarianism, progressivism and the in-built attractions of the new source of power finally vanquished 'backward-looking' environmentalism is less significant to the historian than the process of struggle itself.

Each of these conflicts, together with the controversy over Battersea Power Station, is described in detail in Part II. A major aim has been to identify interactions between social structure and policy formation at the micro level and to emphasise the importance – at times, the primacy – of 'small-scale politics'. A second objective has been to relate social to environmental interests and to hold the ring between those who protested against the 'march of the pylons' and progressives who argued that the urban working class could only benefit from technological advance if the tender consciences of preservationists were ignored. Attention has also been given to the meaning of electricity in different geographical and cultural settings. Thus in London in 1929 during the Battersea crisis, those who opposed the construction of a 'super-station' saw the new form of energy not merely as a threat to health and environment but as an alien presence which would 'corrode' the entire social fabric; in Scotland, English exploitation of Highland loch water was perceived as an affront to the Scottish nation.

A final objective has been to define the conditions under which individual localities were able to organise, either fleetingly or for longer periods, against the full *armatorium* of centralised ministerial power. The struggle, in that sense, was as closely related to the cohesiveness or fragmentation of local communities as to the viability or desirability of large-scale electrical system-building. Given the complexity and diversity of 'anti-pylonism', any attempt to identify an underlying explanation of the intensity and tenacity of the movement, is unlikely to be convincing.

But in a penultimate chapter connections are made between agricultural depression, the emergence of full-blown preservationism, and the putative restoration of a traditional patrician rural order. The chosen method is a close reading of the writings of four influential environmental thinkers – G. M. Trevelyan, Vaughan Cornish, Clough Williams Ellis and Patrick Abercrombie. Together they covered a spectrum of subtly differing ideological positions, running from nature mysticism through romantic regressivism to partial accommodation. If electrical triumphalism spawned numerous and, on occasion, genuinely eccentric variants, rural preservationism was, as we shall see, no less complex and diverse.

# Part I

# Adoption

# 1

# The meanings of triumphalism

Shortages of raw materials and doubts about the shape of the national framework within which individual undertakings would in future operate might well have dampened the spirits of the British electrical industry in the aftermath of World War I. But, in terms of rhetoric and self-image, the predominant mood was one of passionate evangelicalism – in that sense, at least, this was certainly the 'golden age of electricity'.[1] Restricted thus far in its range of industrial and domestic uses, and purchased mainly by the relatively wealthy, the new form of energy was now coming to be perceived as a mass 'service' which would lead to social transformation and constitute a new and dynamic sector within the national economy.

This proselytising optimism was repeatedly tested and undermined by foreign competition, fluctuations in the world economy, and the failure, at crucial junctures, of the industry to put its own collective house in order. But this is less significant than the continuity and intensity of the discourse itself. The universal potential of the 'power of electricity' and of the 'electrical idea' were deployed as a rhetoric and legitimation to combat the antagonism and doubts of 'non-believers'. Old-fashioned 'humanists' who questioned whether 'science' would (or should) remould the world, rural preservationist groups who resisted the spread of pylons, or the self-interested gas and coal industries – all were subjected to 'propaganda', hectoring editorials, pamphlets and films which demanded that they see the error of their ways and yield to the inevitability of 'progress'. 'Doubters from within' were also subjected to criticism. Engineers, salesmen and demonstrators of electrical appliances were adjured, at Christmas and the beginning of the new commercial year, to take stock and remind themselves that the success of the industry, and the quasi-miraculous science on which it was based, depended on their creative

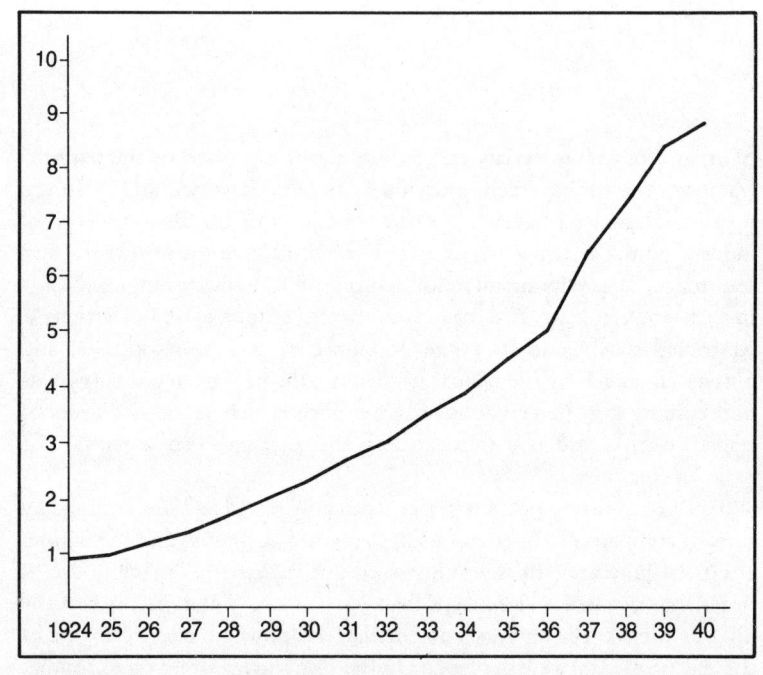

**Figure 2** Domestic electricity supply in Britain, 1924–40 (in millions of units). Data before 1924 is unreliable. 'Domestic supply' covers 'lighting, heating and cooking' but not power used for appliances like hoovers and refrigerators. Source: *Annual Reports* of the Electricity Commissioners.

enthusiasm, mental flexibility and physical fitness. 'Electrical men' in the inter-war years were urged to see themselves as part of a life-enhancing mission. Those who dropped by the wayside – the greatest sinners of all were electrical salesmen in the 1920s who continued to use gas – were as harshly treated as the 'enemy without'. This overwhelmingly triumphalist ideology was itself sustained by subsidiary rationales which locked the new rhetoric into existing cultural concerns.

Popular history, 'futurology' and the crisis of industrial capitalism were drawn upon to point more forcefully to the world-transforming potential of electricity.[2] 'The further extension of electric service', it was argued in a pamphlet of 1920, 'inspires the hope of recovering the arcadian conditions which existed before the age of steam'.[3] The new source of energy constituted in this sense an 'atoner for the undoubted evil of "industrialization" in the Victorian era. The steam engine and the coal-fed boiler made of many fair counties what they are today. Electricity has come to redeem them ... Ten or twenty years hence its complex and as yet unforeseen effects will be examined by every art critic and writer on social welfare.'[4] 'Electricity as atoner' was a theme which would be taken up by politicians, planners, rural propagandists, electrical engineers and journalists. Pylons and low-tension wires would, according to this view, stimulate a second and cleaner industrial revolution, the decentralisation of mass production and a rejuvenation of a deeply depressed agrarian society.

Reevaluation of the past led to speculation about the future. In electrical terms, agriculture was the most retarded sector of all in interwar Britain, but it generated numerous unambiguously utopian visions. In the anonymous 'A Vision of 2024 AD', written in 1924, the author constructed a scenario in which 'electricity it is that garners the crops, dries the hay and corn no matter what the weather may be, milks the cows, warms the house, cooks the food, ploughs the land, makes poultry-keeping pay, allows the labourer to work in comfort and pride'.[5] When they turned their attention back to the industrial and social troubles of the 1920s, and more specifically, to incipient mass unemployment, passionate upholders of the 'electrical idea' were no less sanguine. 'Are we downhearted?' an editorial writer asked in 1924. 'No! Electricity is in our blood!'[6] As for the question, 'Will Electricity Save England?', the answer, despite qualifications, was 'yes, it will'. 'No single agency can "save" us, not even electricity. But electricity, coupled with industrial peace and the will to produce, can bring about a very good imitation of the golden age.'[7]

By 1926, the year in which the government announced its intention to construct the National Grid, enthusiasm for domestic as well as industrial electricity was rampant. The supply industry may have been held back during the National Strike but it was impossible to 'dam back the St Lawrence for ever, any more than one can dam back the patient miners longing for their work. Once the torrent makes a breach there will be a deluge.'[8] The press, and particularly the *Daily Express*, speculated on the miracles of an 'all-electric' society. 'Cooking, heating, washing, drying and cleaning – all these services will be electric when the day of the [farthing] electrical unit is reached. The work of the housewife will be reduced by more than 50 per cent.'[9] Parliamentarians picked up on the theme of electricity as a labour-saving 'social service'. Echoing Émile Zola, a member asserted in 1929 that 'electricity today is as much a national and domestic necessity as are food and water'.[10] The statement that 'if electricity in its present form had been known to the Romans . . . it would have been deified'[11] was patently absurd but it expressed the degree to which the 'propaganda' of the 1920s had insinuated itself into popular consciousness. Cinema, literature and poetry now drew increasingly on electrical themes and terminology, and to Spender's and Auden's appropriation of the imagery of the Grid we should add a largely anonymous body of verse which sang the praises of the new energy source and made play with copper, watts and ohms.

> Shocking!
>
> '*Watt* a girl! I'd like to *copper*',
> The *live-wire* electrician said;
> 'I'd build ourselves a nice big *ohm*,
> And we'd *spark* till we were *dead*'.[12]
>
> Anon
>
> There are cookers large and small,
> Such a benefit to all!
> Cakes and pastries bound to rise,
> Consider – and Electrogize!'[13]

Connections were also made between the 'magic' of the new source of energy and the still vibrant glories of Empire. In 1923, Sir Hugo Hirst, chairman of General Electric, had insisted that 'electricity, like nothing else, provides the means to knit together our far-flung Empire, to the homes of all, by keeping those in the distant farms, in the bush, on the veldt, or in . . . our Overseas Dominions in the closest possible touch with the centre and the heart of the Empire'.[14] By 1937, the year of the Coronation, imagery of this type had been intensified and domesticated.

> The Monarch who has just been crowned before the Nobles of the world will go by himself to a quiet room in Buckingham Palace and speak to the people he has promised to serve. Radio sets in our homes will bring his friendly voice to us; radio sets in the jungle will bring it to his dark-faced subjects; radio sets in the Arctic will bring it to those men who keep his Law in that snow-bound land; radio sets on the ships will bring it to his Navy on the high seas, and radio sets will bring it to his subjects travelling above the clouds in aeroplanes.[15]

The industry had long prided itself on its export of materials and expertise to every corner of the earth, and trade associations devoted entire sessions of their annual conferences to imperial developments. But the exceptionally rapid growth of radio in the 1930s was held to have brought the monarch into more intimate contact with the Empire and to have bestowed reflected glory on those who had made such progress possible. 'Electrical engineers', enthused *The Electrician*, 'can indeed feel a deep sense of pride and responsibility in an achievement which has cemented the Empire in a way that nothing else could have done'.[16]

But the very intensity of electrical triumphalism in Britain during the inter-war period was itself partly created and fuelled by the social and technical problems inherent in the mass adoption of the new source of energy. It was especially difficult to counter the claim, often made by the gas and coal industries and by prospective purchasers, that electricity was exceptionally dangerous. As influential a figure as S. E. Britton, chief engineer at Chester, might provide seemingly irrefutable figures in 1926 to show that electric shock was much less frequently encountered than domestic burns or road accidents.[17] But the very novelty of electrical injuries and the way in which they were exploited by the popular press embarrassed the triumphalists. If only the public, it was lamented in 1920, could be convinced that 'the bath . . . is the one spot in the household where the conditions are highly favourable to shock'.[18]

Yet individuals in the 1920s seemed capable of getting themselves into every kind of electrical trouble. They climbed poles carrying current from collieries; picked up inadequately insulated kettles while wearing faulty radio headphones; and indulged in all sorts of dangerous activity in close proximity to the sink or bath.[19] Nineteen twenty-eight seems to have been a particularly bad year.

> A woman while cleaning the lamp in her kitchen, stood in the sink and held the lamp in the desired position by means of its holder. The insulation of the flexible had perished somewhat, as a result of the steam and fumes usually associated with

the average kitchen, with the result that the woman came into contact with the wire, and, by reason of standing in the partially wet sink, received a severe shock.[20]

In that same twelve months there were widely publicised cases of a boy dying from electric shock in a chip-shop; a housewife suffering electrocution while using a suction cleaner;[21] and the death of a child in Hull following X-ray treatment.[22] Fatalities seemed to decline with the dawning of the new decade – perhaps appliances had become more reliable and the public better educated – but in 1932 the electrical press contained reports of the case of a woman who had killed herself while using an electric hair-drier in the bath.[23] Were there no limits to human folly?

Much of the blame, according to the triumphalists, lay with the popular press, which trivialised and sensationalised, not only the 'new electrical age', but all forms of scientific activity. The press, the *Electrical Review* insisted, had a duty to 'inculcate in the public mind that knowledge and love of science through which "our men of science" – unexcelled in the world – acquired their equipment for winning the war, instead of perpetuating the silly and antiquated notion that they are habitually immersed in useless hobbies of no practical utility'.[24] But little could be achieved in an environment in which popular journalism looked on 'electrical phenomena as being fit matter, on the one hand, for inclusion in the realm of Utopia, and, on the other hand, as part of the weird policies of the kingdom of Laputa';[25] or in which the headline 'METAL TURNED INTO GAS. Age-Long Dream of Alchemists Solved. Terrific Heat' claimed massive numbers of readers.[26] Every example of press 'stunt', sensationalism, or deliberate misinformation, electrical zealots insisted, must be immediately corrected. The idea that the 'normal' sea-tide could be harnessed along every British coastline to produce vast supplies of power;[27] that 'large generating stations must be fixed near large streams so that there may be sufficient water to cool the cables';[28] that, high-tension wires might cause an unnaturally high absorption of current and thus precipitate large-scale outbreaks of 'nerves' (or climatic and digestive disequilibrium)[29] – all such myths must be ruthlessly countered.

So, also, must press accounts of entertainments which concentrated exclusively on the quackish rather than functional aspects of electrical development. Grindell Matthews, who caused a minor stir in London in 1928 with his 'death ray', might be 'a fine figure of a man' who was able to handle 'switches and mysterious levers' in such a way as to hold a large audience ('60 per cent "flappers"') spellbound.[30] But such antics detracted from the social legitimacy of the new form of power and did nothing to raise levels of electrical consciousness. Although 'stunts' and journalistically induced sensationalism declined as domestic supplies were extended to a growing number of urban areas in the early 1930s, triumphalists showed concern over the public understanding of the new energy source throughout the inter-war years.

The EDA had complained in 1920 that 'it is only too apparent . . . that the public have but little knowledge of the economic possibilities and limitations of Electricity Supply'[31] and, in the same year, *The Electrician* estimated that '90 per cent of the population' did not yet realise the utility of the new form of energy.[32] Triumphalist doctrine was predicted on the belief that electrical consciousness would only be significantly raised if the 'unconverted' became familiar both with the basic 'facts' of the science and with those heroic figures, from Faraday to de Ferranti, who had banished ignorance and unbelief. There continued to be demands in the late 1930s, therefore, for more pedagogically sophisticated and co-ordinated programmes for the inculcation of basic electrical principles. 'To produce an electrically-minded generation', wrote a contributor to *The Electrical Age* in 1938, 'the approach has to be two-fold. There must be teaching of electricity on the one hand, and the example of electrical appliances in use at the school on the other'.[33] Electrical and scientific 'awareness' was, and still is, a difficult entity to define and measure, but when in 1939 he surveyed the extent to which triumphalist propaganda had changed public attitudes towards the new energy source, A. C. Cramb, a former director of the EDA, concluded that 'it is not difficult to give evidence of the extent to which the public is still in the dark as to the electric service available to them' and that too many housewives were still afraid of electric shock.[34] Part, then, of the fervour inherent in triumphalist rhetoric can be explained in terms of an over-reaction to the ignorance and indifference of 'non-electric' groups. But another factor making for the intensity of the language which expressed the pro-electrical ethos of the supply industry and its official organs during these years was the extent to which adherents to the 'electrical idea' believed that they could identify 'treachery within'. The principal target here was the salesman who was either too snobbish or too lazy to make the transition from a home which was heated and lit by gas to one which conformed to the new ideal of the 'all-' or 'part-' electric house. There could be little excuse for the company employee who deliberately shunned the product to which he ought to be wholeheartedly committed; and it was for this reason that the early 1920s witnessed the publication of 'homilies' depicting the punishment and salvation of the recalcitrant electrical salesman.

'The Awakening of Peterkin' told the story of a traveller who, in spite of the wholehearted support of his 'pro-electric' wife, had lost all belief in the commodity he was supposed to be selling. Deliberately wasting time, rather than making appointments with clients, the hero (or anti-hero) wanders into an expensive French restaurant, with the intention of blowing what little cash he still possesses on a slap-up lunch. But fate intervenes – the restaurant's antiquated heating and cooking system collapses and the hysterical *patron* bewails that he has no option but to cancel the full-dress dinner which had been booked by a large party for the same evening. The crisis goads the stick-in-the-mud salesman into action. He clinches the order and personally supervises the super-fast installation of cookers, heaters and more effective lighting. The *patron* treats him to a meal and, on his way home, the salesman orders a range of electrical appliances for himself. An excited servant informs his wife that 'there's a toaster, mum, in the dining room, and a cleaner for the carpets, and a hot water thing . . . and a radiator in the bath-room and a lot more. The men what was here went at it as though they were crazy – they worked that hard. I've tried 'em all – and they all work.' Then the salesman himself returns. ' "John", she whispered unsteadily, "I hadn't the faintest idea of all this – but I believe – ". He took her in a sudden bear-like hug. "That's exactly it, Marjory – for at last I believe too." ' Mills and Boon combines with electrical triumphalism to provide an uplifting and salutary climax.[35] To these often imaginative experiments in conversion were added the more forthright accusations of the electrical press. 'The only good point', *The Electrical Review* inveighed in 1923, 'that can be advanced by way of excuse for their continuing to consume gas is the valuable practical experience that they derive from being customers of a gas company with all the petty annoyances, shortcomings, escapes, smells, explosions and what-nots, that are the penalty!'[36] The salesman who refused to convert to electricity was 'the stick-in-the-mud type permanently within our gates'.[37] The worst of the crisis appears to have been over by the later 1920s, with the adoption of an 'all-' or 'part-' electric house now evidently becoming less socially experimental and potentially stigmatising. But salesmen continued to be chided for their failure to demonstrate absolute belief in their product and they were informed, in 1927, that 'if everyone in the electrical industry would make it a point to take every available opportunity to assist in spreading what has come to be called the "Electrical Idea" the results would be astonishing'. Far too many company employees, evidently, were still reluctant to talk shop outside office hours.[38]

The plausibility of the ideology of electrical triumphalism in inter-war Britain depended on the degree to which it could draw on existing cultural repertoires while simultaneously generating novel images of technological superiority, cultural modernity, and near-universal access. The context which overwhelmingly determined the working out of these claims and which further illustrates the distinctive intensity of triumphalism itself was one in which a much older form of energy – gas – still held the whip-hand. The aggressive evangelicalism which lay at the heart of pro-electric propaganda can, in this sense, be conceived as a defensive reaction to the cultural dominance of the older utility. This defensiveness expressed itself from the very beginning in savage assaults on the alleged dirtiness and danger of electricity's principal rival. 'Coal and gas', it was argued in 1920, 'cannot by any possibility render service equivalent to that of electricity, but they can, and do, introduce an intolerable amount of dirt, dust, labour, and pollution of the atmosphere, all of which have to be paid for in terms of health and pocket'.[39] Countering accusations that electricity caused large numbers of domestic accidents, pro-electric apologists ghoulishly argued that 'corpses [as a result of gas explosions] are so plentiful that even without the newspapers every little community sooner or later has first or second hand evidence of the truth'.[40] These images of an industry which facilitated or actually encouraged accidental death were sometimes taken to outrageous extremes but emphasis was also laid on the environmental deficiencies of the older industry.

Like many another minister during these years Neville Chamberlain was interpreted as offering tacit state support to the electrical industry when he spoke in 1923 of the clearness of urban skies during the 'great coal stoppage'.[41] This was a theme which the triumphalists took up and developed with gusto, and, although they might frequently blame coal and the 'antiquated hearth' for atmospheric pollution and the omnipresent 'London Particular', gas was never far from their sights. London, it was asserted again in 1923, was not 'naturally a foggy place but one that has been made so by human agency. The moral of all this is clear enough: the reduction of smoke-producing grates, furnaces and fires, and the substitution therefor of something that will give us an atmosphere of which even the country will be envious.'[42] But the new form of energy must not be a fuel which, in addition to causing numerous explosions and spreading dust throughout the house, also made a massive contribution to 'pea-soupers'. 'Smoke', *The Electrician* claimed in 1924, cost the 'approved societies in 1923 £2,000,000 in sickness benefit and the nation £3,000,000 for insured persons alone'. This, it concluded, was 'almost, if not quite, as bad as strikes'.[43] Associations were also made between smog, crime and suicide[44] and, more credibly, in the light of hypotheses put forward by J. B. S. Haldane, between a polluted urban atmosphere and cancer.[45] Pro-electric ideologists insisted that it would be unrealistic to expect any sudden, overnight replacement of 'primitive' and environmentally harmful fuels, such as gas and coal, by the 'clean rays' of electricity. But the transitional period between the final eradication of the older domestic fuels and the emergence of a civilisation based on electricity would witness the newer form of energy bestowing enormous curative benefits via its 'natural' healthfulness.

It was a short step from a vision of the innate therapeutic powers of electricity to a conviction that the 'diseases which beset the industrial population owing to the scarcity of real sunlight may be arrested, and in time wholly eradicated, by the artificially produced article'.[46] By now there was a corpus of solidly researched scientific work which indicated that 'supplementary sunlight' could undo some of the worst effects of deficiency diseases, and particularly rickets. But triumphalists extrapolated massively from this medically authenticated position and presented 'sunlight' treatment as something which could cure, beautify, and protect in an environment plagued by the unnatural effusions of the older and dirtier industries. Although it was frowned on by responsible medical men this ultraviolet cult threw up a weird diversity of experience and experimentation. Slum children paid regular visits to municipal 'sunlight' clinics; rheumatics treated themselves to over-extended 'sunbaths'; middle-class children in pyjamas and goggles were given daily doses of 'health'; and their mothers indulged in dangerously lengthy 'beautification' sessions.[47]

Given the continued relative decline of coal as a domestic fuel there seemed good reason, from the 1920s, for there to be increased and 'active co-ordination between the two great sister industries of lighting and power, namely, the gas and electrical industries'.[48] But attempts to devise a state fuel policy during the 1920s invariably foundered on the mutual suspicion which plagued relations between the two industries. The pro-electric lobby was aware – and after the beginning of the construction of the Grid convinced – that Conservative and Labour governments alike would champion and invest heavily in electricity as the energy source of the future. For their part, leaders of the gas industry railed against this in-built bias, modernised their product and appliances and stabilised their costs.[49] Tension and mutual recrimination mounted. When co-ordination between the two industries was again mooted in 1926, the *Electrical Times* personified gas as a 'would-be bride . . . a trifle long in the tooth . . . the object of her affections is that exceedingly virile youngster, electricity. Naturally, at her time of life she would like to feel settled.'[50]

By the early 1930s bearers of the electrical standard were hitting hard and low. 'The gas industry's turnover in cash and corpses is a world wonder. They consider it also a very popular industry and it certainly is the most popular form of suicide, many hundreds die of it every year, but it is painless and that is in its favour.'[51] 'Coal and gas emanations at full blast' were claimed to have 'tipped the balance for sufferers from bronchial and other winter infections. Thus have many thousands been shuffled off this mortal coil, duly certified by the doctors as dying from natural causes.'[52] As the gas industry improved the design of its appliances, and more than held its own in the domestic sector, triumphalist attacks intensified. 'Less than half the heat units in coal gas are effective', the *Electrical Times* claimed in 1931; gas 'emits hot dangerous fumes which require a chimney . . . even with a chimney in use much effluvium escapes to poison the surrounding air'.[53] It was hardly surprising that abuse as scabrous as this goaded the gas industry into marking the formal opening of the national electricity network in 1934 with the publication of a stridently anti-electric pamphlet, *The Lure of the Grid*, or that the ever-sensitive pro-electric camp should have responded in 1935 that 'the increasingly anti-electrical advertising and publicity used by many gas undertakings leaves no doubt that warlike tactics are deliberately being adopted'.[54] The battle continued until the outbreak of hostilities and well on into the post-war period.

British governments throughout the inter-war years frequently acted as though they believed electricity to be the 'energy of the future'. But they were unwilling to introduce overt market discriminations against rival industries. This would have been politically unwise and might have involved further destabilisation of an already chronically unstable economy. Yet the claim of those most vehemently opposed to the mass adoption of electricity that the very existence of the Central Electricity Board represented a blatant ideological and economic subsidy to the new source of power was a valid one: when pro-electric triumphalists hectored and lectured in the name of science and modernity, their message was undoubtedly enhanced by the Board's backing and high political status. One needs to distinguish here between the scientific rhetoric of electrical triumphalism and the more muted, bureaucratic style of the Board's executives and spokesmen, but the cause of electricity was greatly strengthened by the interaction of the two discourses, and of the aims and strategies embedded in each of them. Electrical triumphalism was more than a generalised and at times quasi-religious rhetoric, growing out of and moulded by a deep defensive reaction to the existing technological order. It must also be seen as the shaper and bearer of images which made up the 'public face' of electricity and which persuaded large numbers of the uncommitted to invest, symbolically as much as economically, in the 'future' rather than the 'past'. The task of persuasion – the conversion of the raw passion of triumphalism into culturally enticing publicity – was taken up by the Electrical Development Association.

**Notes**

1  The first use of the phrase that I have come across is by Sir Henry Renwick in the *Electrical Review* (hereafter *ER*), 6 October 1922, 471.

2  In his provocative *Objects of Desire: Design and Society 1750–1980* (1986), Richard Forty coins the phrase 'millenarian' to describe intense electrical progressivism. My reason for preferring 'triumphalism' is that 'millenarianism' is already established in several important, non-technological historiographical settings. For further context on inter-war scientism and progressivism see Roy and Kay MacLeod, 'The Social Relations of Science and Technology, 1914–1939' in Carlo M. Cipolla (ed.), *The Fontana Economic History of Europe: the Twentieth Century: Vol. 1* (1976), 301–63; and Gary Werskey, *The Visible College: a Collective Biography of British Scientists and Socialists of the 1930s* (1978).

3  'The World's Right Hand', Electrical Development Association (EDA) pamphlet 109 (1920), 11.

4  *Electrical Times* (hereafter *ET*), 8 March 1923, 234.

5  'A Vision of 2024 AD', EDA 479 (1924).

6  *ER*, 5 December 1924, 842.

7  'Will Electricity Save England?', *ET*, 3 December 1925, 669. See, also, in similar vein *The Electrician* (hereafter *E*), 29 May 1925, 619.

8  *ET*, 7 October 1926, 407.

9  *Daily Express*, 20 August 1927.

10  *Hansard* 233, 11 December 1929, 500, W. Wellock, member for Stourbridge. See, also, the comments of D. Hopkin, member for Carmarthen, during the same debate, 495–6.

11  'Electricity and the Public', *E*, 28 April 1930, 693.

12  Nicholas Manula, 'Shocking!' *Electro-Farming* (hereafter *EF*), August 1928, 70.

13  *Rural Electrification and Electro-Farming* (Hereafter *RE-EF*), January 1930, 230.

14  'A Broadcast Story of the Electrical Industry', EDA 405 (1923), 7–8.

15  'May 12th. A Souvenir for Children Commemorating the Coronation of

King George VI and Queen Elizabeth', EDA (1937), 25.
16   E, 21 May 1937, 666. See, also, Phyllis Perkins, 'Electricity – the Girdle of the World', *The Electrical Age* (hereafter *EA*), Winter–Spring, 1939, 515. For the massive cultural impact of radio see Asa Briggs, *History of Broadcasting in the United Kingdom: The Golden Age of Broadcasting* (1965).
17   S. E. Britton, 'Supplies to Outlying Districts', *Proceedings of the Incorporated Municipal Electrical Association* (hereafter *PIMEA*), 1926, 29.
18   ER, 16 April 1920, 482.
19   *Hansard*, 9 May 1923, 2396. Major F. Kelley, member for Rotherham, and ER, 19 September 1924, 418.
20   EA, 6 April 1928. See, also, the comments of Sir M. E. Manningham-Buller, member for Kettering, in *Hansard*, 223, 11 December 1928, 1901.
21   E, 27 July 1928, 87.
22   E, 28 December 1928, 727.
23   E, 1 April 1932, 466.
24   ER, 27 August 1920, 259.
25   E, 30 September 1921, 401.
26   ER, 31 March 1922, 435.
27   ET, 31 July 1924, 1930.
28   ET, 4 February 1926, 134.
29   Cited in ET, 21 November 1929, 866.
30   ET, 7 August 1928, 156.
31   'The British Electrical Development Association (Inc)', EDA 76 (1920), 3.
32   'Coordination the Keynote', E, 24 September 1920, 347.
33   Frederic Evans, 'Making Children Electrically Minded', *EA*, Summer 1938, 429.
34   Alex C. Cramb, 'Advertising and Selling Electric Service', *PIMEA*, 1939, 125.
35   'The Awakening of Peterkin', EDA 191 (1921). The anonymous writer was Alan Sullivan – EDA Council *Minutes*, 21 October 1921. For a sceptical review see E, 19 August 1921, 245.
36   ER, 6 July 1923, 261.
37   'On Making Progress', ER, 28 December 1923, 961.
38   E, 20 May 1927, 542.
39   ER, 11 June 1920, 738.
40   ET, 26 January 1922, 74. See, also, 'Helping Smoke Prevention', ER, 20 April 1923, 601.
41   ET, 5 July 1923, 17.
42   E, 28 December 1923, 717. For other contributions to the long-running 'smoke' and 'hearth' controversy see ER, 21 December 1923, 923; Herbert M. Berry, 'Electrified Houses of Great Britain with Special Reference to the Heating Question', *PIMEA*, 1927, 196–7 and ET, 1 January 1931, 2.
43   E, 21 November 1924, 575.
44   Mrs H. A. Howie, 'Electricity and Health', *PIMEA*, 1925, 111 *passim*.
45   E, 30 January 1925, 112–13.
46   E, 19 August 1927, 218.
47   The literature on this subject is voluminous. But see N. Bishop Harman,

'Light and Health', *E*, 8 April 1921; R. Rook-Jones, 'Ultra-Violet Rays and Health', *EAW*, October 1927, 221; 'The Use of Ultra-Violet Rays', EDA 915 (n.d., probably 1929); 'Ultra Violet Light Treatment: What of the Future?', *REEF*, August 1929, 84; and Elizabeth Sloan Chesser, 'Electricity Applied to Health', *EA*, April 1936, 57.

**48** *Hansard*, 179, 9 December 1924, 145. R. G. Clarry, member for Newport. See, also, *E*, 30 October 1925, 495.

**49** On gas prices see Leslie Hannah, *Electricity Before Nationalisation: a Study of the Development of the Electricity Supply Industry in Britain to 1948 (1979), 206*.

**50** *ET*, 7 October 1926, 407. For similar opposition to 'fusion' between the two utilities see ibid, 22 November 1928, 709.

**51** *ET*, 12 February 1931, 280; see, also, *Electrical Industries and Investments* (hereafter *EII*), 21 September 1932, 1478.

**52** *ET*, 5 March 1931, 422.

**53** Ibid, 26 November 1931, 838.

**54** V. W. Dale, 'Six Reasons why the Industry should Support the Million New Consumers Campaign', EDA Handout, 5 December 1935. For a warning that the 'struggle' will be long and bitter see 'Is the Grid Justified?', *E*, 8 March 1935, 96 and on the *Lure of the Grid*, EDA Council *Minutes*, 19 January 1934.

# 2

# Constructing the image

Founded in 1919 as an offshoot of the Heating and Cooking Committee of the Institution of Electrical Engineers, and funded through contributions from supply undertakings and major trade associations, the EDA worked with a small staff and a meagre budget under the directorship of an able ex-municipal engineer, J. W. Beauchamp. At first the new organisation strove to establish the validity of 'collective' advertising for public utilities and to counteract the impression that it wished to represent or push the products of any single producer or sector within the industry as a whole. This relatively new form of industry-wide advertising had to be carried on in an environment of deep public ignorance. 'Most people', the EDA claimed in 1920, 'have but little knowledge of the economic possibilities and limitations of Electricity supply'.[1] A year later, the *Electrical Times*, deploying characteristically pedagogical imagery, insisted that 'the public are waiting willing to be educated in electrical lore ... EDA is to be the teacher, and now is the time to equip it for its onerous duties.'[2]

But the 'equipping' of the industry's major educative and image-building agency proved to be a stumbling-block throughout the inter-war period. The private supply companies failed to contribute on a sufficiently generous scale, with the result that staff found themselves spending valuable time soliciting for funds rather than devising compelling campaigns.[3] This played into the hands of the British Gas Association (BGA) and did little to convert uncommitted undertakings to the idea that collective publicity would generate higher levels of demand. By 1924, there was widespread feeling that the organisation needed more generous funding.[4] In 1925 the EDA introduced reforms which required supply undertakings to donate one-tenth, rather than one-fortieth, of their annual revenue from the sale of electricity.[5] The immediate response in the trade press and the industry was favourable. 'There has been a generous response to Eda's appeal for stronger support', an anonymous observer noted, 'the Association is more than double as strong as it was, it is now a solidly established and permanent machine for developing the uses of electricity.'[6] The Association was also said to have 'won its spurs, it has become an established focus for the collective inoculation of the public with the electrical idea'.[7] But by 1928 old fears and criticisms were resurfacing. The organisation had certainly co-ordinated several well-designed campaigns and strengthened its provincial base via the establishment of regional committees. But its press work seemed weak when compared to that of the BGA. It still lacked the full 'moral' support of the industry. And, most debilitating of all, many of the more conservative supply undertakings had yet to take out membership.[8] At the annual general meeting in 1928 the president, R. P. Sloan, made a desperate plea for increased support. 'He strongly urged those who did not subscribe their full quota, or who had given what they had given grudgingly, or had not given anything at all, to give the matter serious consideration and assist in carrying on the good work.'[9] In 1929 an accusatory motto was appended to the back page of the report of the annual meeting. 'A past President of the Association truly said – "that those Supply Undertakings and Firms in the Industry who are not members of the Association, or who are not making an adequate contribution to the funds, are really reaping benefits which have been paid for by others".'[10] Moral persuasion could do so much, but no more. Compared with the highly financially centralised gas industry, national electricity supply was still fragmented. The construction of the Grid would in time compel the state to play a more centralising and rationalising role, but, now, in the late 1920s, it was still possible for many of the smaller, as well as a minority of the larger, concerns to concentrate on long-established outlets and pay scant regard to growing latent demand for domestic electricity. By 1930 the EDA was receiving little more than £30,000 from the supply industry – less than one-third of the amount spent on 'co-operative' advertising by the 'gas interest'.[11] This was in an environment in which 'many hundreds of people still visualise even electric lighting as a luxury and an expense'.[12]

When A. C. Cramb, another experienced ex-municipal electrical engineer, became director in 1930, he faced an exceptionally difficult task: to triple the contribution of the supply companies and devise a sales and advertising policy to match the widely acknowledged seductiveness of the gas industry's well-established publicity logo, Mr Therm.[13] Cramb's failure to rejuvenate the EDA is partly explicable in terms of the growing impatience of the CEB and its fear that greatly increased supplies of electricity available on completion of the Grid would not be matched by a sufficiently buoyant demand. Cramb was also stymied by the continuing conservatism of the supply undertakings. Many of these, it was lamented in 1931, 'cannot or will not do their share in providing the motive power to increase the amplitude of EDA's radiation. Is senile somnolence or wilful blindness the cause?'[14] Sir Andrew Duncan and Archibald Page at the CEB were in no mood to wait around for answers to rhetorical questions and they now encouraged the General Purposes and Finance Committee of the EDA to draw up a blueprint which would allow electricity to wage a tough and uncompromising 'propaganda' war against gas. The Committee concluded that the financial 'deficiency has been due to the failure of the various sections of the electrical industry to respond in full to the appeals for support made by the Council of the Association'. This, together with the imminent completion of the Grid, necessitated an unprecedentedly aggressive sales-drive, co-ordinated by a restructured advertising and public relations agency.[15] Radical recommendations were brought before an extraordinary general meeting in late July 1933. EDA membership was to be restricted to 'authorised undertakers' and manufacturers were to be excluded. The CEB would be formally affiliated, make a large annual subscription, and match suppliers' contributions pound-for-pound.[16] Organisational change was reinforced by changes in personnel: Sir William Ray, leader of the Municipal Reform Party and a long-standing member of the London County Council, was appointed to the new position of executive director and Cramb was demoted.[17] The EDA had been transformed from a voluntaristic body into a public relations bureaucracy umbilically linked to the CEB in the post-Grid era.

The influence of the Board was felt in areas as diverse (and trivial) as choice of agency for advertising campaigns and the checking of proofs.[18] But it was in the sphere of finance and revenue that the change initially seemed to promise most – by 1934 subscriptions had risen to approximately £85,000, with the CEB paying £42,000 and the municipal undertakings contributing £31,000.[19] Now, following two years 'hamstrung by uncertainty', the reconstituted body found itself confronted once more by the full fury of the 'gas offensive'.[20] The Association was urged by the trade press to refute the charge that 'most of the electrical propaganda hitherto carried on has been designed to appeal to a more prosperous class of consumer' and to improve the quality of its publicity material.²¹ None of these challenges was convincingly met in the years that followed – least of all the need to respond speedily and aggressively to the anti-electric propaganda so expertly produced by the gas industry. 'Mr Therm' had shown himself to be a brilliant and instantly recognisable reminder of the virtues and omnipresence of the older utility, and electricity never succeeded in creating a convincing rival. ('I'm Electric' was a blob-like, sub-Disney creature, 'vastly derivative of the Mr Therm campaign' and, in visual terms, wholly unrelated to the new form of energy.²²) But the persuasiveness of 'propaganda' material during this period must also be related to institutional variables – the business acumen and flexibility of the reconstructed EDA Council; the creativity of permanent and free-lance staff; and, finally and crucially, to levels of funding. In the bright new dawn of 1933, the CEB had hoped for an increase in membership which would have raised annual contributions to £100,000 – making a grand total of £200,000, after 'topping-up'. But many smaller companies were still suspicious of the benefits of collective publicity and, in 1936, the figure stood at little more than £45,000; it rose only slowly during the rest of the decade.²³

The most revealing index of all was a comparison published in 1939 of press advertising between electricity and a selection of 'new' and consumer industries (many of which, ironically, depended upon electricity either for manufacture or operation). Statistics such as those presented in Table 1 reinforce the impression that electricity was significantly 'under-advertised' throughout the inter-war years and that higher investment in 'sales' would probably have made greater inroads into territory traditionally dominated by gas.²⁴ A distinction needs to be made between 'official' EDA publicity – centrally produced pamphlets, posters, hand-outs and films systematically targeted at suppliers, manufacturers and consumers – and local advertising and word-of-mouth information disseminated by concerns which were not EDA members. It is exceptionally difficult to undertake accurate quantification in this field but the poor financial performance of EDA between 1919 and 1939 strongly implies that in national terms 'non-official' and uncoordinated publicity predominated.

Despite its failure to rally the smaller suppliers to the electrical cause, or to seize the high ground from gas, the Association produced an extraordinarily large and varied body of publicity material; the benefits derived by non-members from this output should not be underestimated and were, indeed, repeatedly commented upon by those who were seeking to boost revenue for the metropolitan organisation.²⁵ The most

**Table 1** *Comparative Press Advertising c.1938*

|  | £ million |
|---|---|
| Motors and cycles | 2.1 |
| Patent medicine | 2.6 |
| Beverages | 1.9 |
| Beauty aids | 2.1 |
| Cigarettes and tobacco | 1.8 |
| Furniture | 1.0 |
| Radio | 0.7 |
| Electricity | 0.1 |

*Source*: A. C. Cramb, 'Advertising and Selling Electric Service', *PIMEA*, 1939, 146.

impressive of the EDA's work was undoubtedly more sophisticated than all but a fraction of the local material.[26] But did writers, illustrators and film directors play anything more than a marginal role in shaping mass perceptions of the 'electrical revolution'? Nearly every short story, instruction book, or film produced during this period either featured or made an unambiguous appeal to the upper-middle or middle-classes. Policy makers were evidently convinced that it made better economic sense to sell an increasingly sophisticated range of appliances to the moneyed classes than attempt to provide the 'basic essentials' to every household in the land. Copy-writers idealised the 'modern', 'leisure-class' life-style and directed it at all those who possessed sufficient disposable income to purchase an extensive range of consumer durables. Viewed as a whole, this body of work incessantly contrasted implied technological upheaval – scientific revolution, almost – with deep social conservatism. Within this cultural context, electricity would certainly 'change the world' but only in an ideological universe predicated on and delimited by traditional evaluations of home, class, gender and social mobility. The new form of energy was depicted as being potentially available to 'the people' but the fact that it was displayed by middle-class 'presenters' in middle-class settings acted as a strong deterrent against any idea that it should be defined or subsidised as a 'social service' or 'free good'. The fact that the first great wave of 'electrical propaganda' immediately after World War I was so closely tied to the 'servant problem' further reinforces this seemingly paradoxical contrast of technological triumphalism and social stasis.

Copy-writers at the EDA in the 1920s capitalised both on an alleged shortage of domestic service and the persuasive idea that, unlike a 'surly' or 'unwilling' maid-of-all-work, electricity would 'never let you down'. 'After a considerable experience of maids', wrote the author of the parable-like 'The Letter of a Householder', in 1919, 'my wife and I held a council of war. She decided to advertise for a Companion-Help who would share with us in every way, and I was to study the elimination of heavy work. *I concluded that electricity was the thing I sought.*'[27] The new source of power was tractable, reliable and subservient, 'a universal servant with an eternal willingness to work'.[28] Girls returning from munitions factories or other types of war-work were claimed to be 'difficult', but 'Miss Electricity' was 'a *Good General Servant* always ready and willing to perform work without bother or fuss. It has been a pleasure to work with someone who never sulked or failed when important occasions arose.'[29] In the bizarre tract of 1921, 'House Burned Down: Dinner as Usual, Doris', it was implied that some servants might just about understand and collaborate with the more reliable and less temperamental form of labour-power. The pamphlet was presented in the form of a chatty letter from the upper-middle-class Doris to her friend Phyllis. Acting on impulse, husband Jack had arranged for the house to be wired, and had stocked it with a veritable treasure trove of appliances. But Doris was unprepared and carelessly dropped a cigarette end with the result that the family home was totally destroyed by fire. In anything other than the brave new world of electricity, catastrophe would have ensued. But Doris, assisted by the aged and untypically reliable maid-of-all-work, Katie, salvaged some of the appliances, and carried them over to a newly built and conveniently located bungalow on the other side of the lawn. At first, Katie is sceptical of the new apparatus but, under the firm guidance of her mistress, she is converted and has an edible meal prepared for the astonished Jack on his return.[30]

The rich were advised to cut back on hard-to-find-and-keep country-house servants and the middle class to do without.[31] In 'The Man who Never Missed the Train' (1924), the insufferably self-satisfied 'Wilkins' points to the inefficiency of servants and lectures a colleague in the following terms: 'There is my talisman. That little switch calls me in the morning, boils my kettle for me, heats my shaving-pot, warms my room, gives me light on a dark day. No waiting on a sleepy maid to bring me tepid water when it pleases her, and give me cold tea. My servant is in the room with me night and day.'[32] Here, and elsewhere, the 'marvel' of electricity licensed an unqualified denunciation of the working and 'serving' classes, with the 'automatism' of the former throwing the independence and alleged unreliability of the latter into sharp relief. The most efficient method of solving the 'servant problem', in other words, was to become servantless. A similar approach – portraying electricity as a supernaturally efficient servant while simultaneously casting aspersions on the class from which domestics had traditionally been drawn – survived into the early 1930s. 'No flue, no smell, no dirty fumes, no matches, no waste of time, no dirty, hot stuffy kitchen. These are not mere advertising catch phrases, they are indisputable points which show you that the electric servant is the best servant.'[33] Servants, then, were inherently feckless and immoral. Better educated than their Victorian predecessors, always likely to be seduced by a widening range of urban leisure pursuits, and well aware of the higher wages and greater degree of independence available to young women working in factories and offices, they no longer conformed to a predictably docile class stereotype. Much better, the EDA continued to urge, to invest in electricity. 'The Servant business is a real problem', reiterated a pamphlet in 1933. 'More difficult to get every day – not often the right sort when you get them ... ELECTRICITY is the cleanest, hardest working, most willing and cheapest servant under the sun. Always on duty, ready for instant service, day and night, at a touch of the switch.'[34]

Copy-writers were evidently aware of the consequences of doing without domestic help, and of seeming to opt out of a class system which placed high status on the ability to pay for and to be seen to be paying for domestic help. It was for this reason that they emphasised the positive advantages – at times, it was argued, the sheer fun and joy – of doing one's own housework by electricity. It was because of this, also, that a distinction came to be made between the 'old', dirty domestic regime, and the new 'modern' order which enabled a housewife to remain clean and healthy, while simultaneously protecting her family from dust and germs. 'An electrified home', the author of 'From Dirt and Darkness to Sweetness and Light' pointed out in 1920, 'is free from sticky and poisonous fumes, is kept clean easily all the year round, and altogether doubles your comfort and halves your work.'[35] 'USE THE ELECTRIC METHOD to clean your house in the Spring', housewives were urged in the same year, 'and keep it healthy all the year round ... With the help of Electricity you can do housework without making more work, and clean the Home without making yourself dirty.'[36] By 1925 'work' with the vacuum cleaner was being described as 'really Play'.[37] But that this 'play' should never be simply enjoyed and must always be directed towards the welfare of husband and children, was reinforced in the later 1920s and early 1930s by advertisements which dwelt with much greater specificity on the relationship between dust and 'germs'. 'Dust carries disease germs', warned 'New Arrivals in the Home' in 1933, 'sweeping merely scatters dirt and dust about the room, makes endless work, and still leaves a great deal of dust in the carpets. Banish dust, dirt and disease by using an ELECTRIC SUCTION CLEANER which lifts all the dirt direct into a sealed bag without a particle escaping.'[38]

Strong associations were also forged, again within the context of exclusively upper-middle and middle-class households, between the 'healthy home' and the 'house beautiful'. Within this idealised sphere, women were encouraged to become increasingly elegant and figure-conscious. 'Iron in Comfort the Electric Way' was dedicated in 1920 to 'the women who take a personal pride in their own dainty wear; the women who do not want to squander youth and beauty'. Physical and technological seductiveness, in other words, interacted with and reinforced one another within the four walls of the 'modern', part- or all-electric house; 'an Electric Iron' was said to be 'a thing of beauty – no other iron is like it – and to continue the narrative, it is a joy for ever to millions of women'.[39] By the end of the decade cultural emphases of this type were largely unchanged. 'There will be beauty in the house you want', a copy-writer rhapsodised in 1929, 'delicate hangings, brightly coloured shades to the lamps, and light coloured paper and paint, as these will last for years in an electric home.'[40] The 'ideal home', reflecting the essentially 'modern' beauty of 'new' middle-class women who had learnt to do without servants, can only have seemed bizarre to those still referred to in electrical publicity as 'artisans'. Bourgeois 'laboratories for living' were to be fitted with three plugs in every room, and candelabras and standard lamps tastefully and 'aesthetically' illuminated; all this at a time when the great majority of working-class homes still lacked even a basic supply.[41].

Even when it was realised, in the early 1930s, that the next focus for intensive publicity should be the 'universal' electric cooker, copy-writers continued to operate within a cultural and institutional climate dominated by the 'modern', independent woman, the 'house beautiful' and the servant-saving capacities of electricity in the larger, bourgeois house. The 'new' woman, it was triumphantly announced in 1932, would be 'free! Having put the dinner on you can forget it. You can go shopping, play a round of tennis or golf, or pay a call, confident that the dinner will be ready, perfectly cooked, when you come home.'[42] The labour expended in the kitchen would be qualitatively different from the drudgery of the past. 'When electricity is installed for all cooking purposes', middle-class housewives were informed in 1933, 'you turn your kitchen into one of the cleanest, neatest and most pleasurable rooms in the house. Cooking by electricity is certain ... simple, healthy and economical.'[43] This image of the kitchen as a 'cooking workshop', with the mistress of the house exercising instant push-button control, had, as we shall see, close connections with ideas derived from the scientific management and time-and-motion-study movements introduced to Britain during the inter-war years by American specialists in 'household science'.[44] Economy and rationalisation of physical movement, greatly diminished amounts of time spent at the stove, certainty that the 'invisible' energy would do its work with absolute consistency and reliability – all this meant that food preparation would become 'an almost automatic process'.[45] But cooking, and the kitchen, would only be finally transformed if the arch-villain, gas were to be totally renounced. Unlike its competitor, the new source of energy, was never affected 'by wind or weather. There are no flues to need sweeping, no fire that won't draw, no days when it cooks too slowly "because the pressure is low".'[46]

In a publicity hand-out, entitled 'I'm Busy Cooking the Lunch', published at the end of the decade and showing an elegant housewife lounging in an armchair, the theme of economy was added to health, hygiene and technical control. 'After the first thrill of having the cooker', the copy ran,

I began to wonder about the expense. One evening I tackled Jack about it. [EDA husbands were often called 'Jack'.] 'Look here', I said, 'can we *afford* to cook by electricity? It certainly makes life much easier and meals much nicer, but I don't want to end my days in the workhouse.' 'You won't have to', he replied; 'it's cheaper than the old way. It only uses about a bob's-worth of current a week – I don't think even your mother would call that extravagant.' Nor did she – in fact, she went and ordered one herself on the strength of the dinner I gave her. Take my tip and get your husband to look in at the electricity showrooms and do likewise.[47]

Two other aids – refrigeration and hot water – would, according to EDA publicity, ensure an even smoother running of the kitchen. The domestic refrigerator had 'started its peaceful penetration into the English home' as early as 1924;[48] and by 1929 the EDA was claiming that the 'three meals a day problem' had been solved by the 'greatest household aid the 20th century has produced ... the British Electrical Refrigerator'. 'The cost of electric current will be small', the same pamphlet continued, 'usually only a few pence a day, with electricity at one penny per unit, while the *money value* of the food saved from spoilage will often prove to be greater than the cost of obtaining and maintaining an electrical refrigerator.'⁴⁹ By the mid-1930s, increased awareness of the dangers of tainted food, as well as the introduction of tougher legislation against excessive use of harmful additives, had persuaded the Association to launch a series of intensive 'cold storage' campaigns. But domestic refrigerators remained expensive, and, compared with America, sales were poor. It was hardly surprising, therefore, that it should have been the upper-middle and middle classes that were once again chosen as the principal targets for official publicity. 'A refrigerator', the social elite was reminded in the mid-1930s, 'saves you and yours from [the] dangers' of illness via infected food.'⁵⁰ Potential purchasers were also given advice on how to store items in the refrigerator – all strong-smelling foods must be covered – and introduced to the skills of 'cold cookery'. Iced consommés, American-style salads and a wider range of desserts could all now be more easily added to the dinner-party menu.'⁵¹

During the hot-water campaigns of the mid- and late 1930s minor concessions finally began to be made to the needs of 'Mr and Mrs Robinson of Acacia Avenue'.'⁵² But, for the most part, it was the image of the smiling, 'modern' *haute bourgeoise* controller of the kitchen which continued to be dominant. Plentiful hot water, the copy-writers now told their clientele, made the man of the house more likely to 'start the day in a better temper'. 'As for the children', a stereotypical middle-class mistress of an all-electric house was depicted as saying, 'I can't get them out of the bathroom when it's their bed-time'.'⁵³ Photographs of the sheer joy, and, on occasion, the erotic potential of electrically heated water, now became more common. Cooking might make heavy demands on the mistress of the house, but 'she's still smiling because washing up is so easy with electric hot water'.'⁵⁴ The controller of the ideal kitchen was presented, then, as effortless toiler, elegant planner, and prescient protector of husband and children. But it was the third of these roles which was most heavily and frequently emphasised and, in this sense, the 'electrical revolution' helped reproduce a successor generation to the Victorian and Edwardian bourgeois wife and mother, restricted almost entirely to the domestic sphere.'⁵⁵

During the 1930s, the 'moral tale', extolling the virtues of the new form of domestic energy, gradually gave way to the visual immediacy of documentary photography: the EDA now tentatively ventured into areas which were as much overtly 'social' as narrowly domestic. The reliability of thermostats in office buildings; the advantages of all-electric flats compared with gas-dependent slums; the pitfalls of poorly placed lights in the 'part-electric' house – text and imagery in material of this type sought to move beyond the 'house beautiful' and to forge links between the electrical industry and builders, architects, town planners and the medical profession.[16]

When talkies were brought into service as part of the industry's propaganda offensive, more predictable themes recurred. In the musical, semi-futuristic fantasy *Plenty of Time for Play*, set in '1955', two upper-middle-class friends gossip in a kitchen about how to prevent a tedious guest from dominating the dinner conversation.[17] The 'over-academic' visitor is brought back by mini-helicopter and before the meal he and the man of the house watch a live outside broadcast of 'Don Bradman Jr' scoring thousands of runs against an English attack of the mid-1950s. In cinematic and dramatic terms, the dinner is predictably flat. The 'boring guest', who just happens to be writing a scholarly work on the 'pre-electric' age is first teased and then 'converted' by the women; anyone who nostalgically looks back to the good old 'non-electric' days is either a fool or an anti-social inhibitor of scientific progress. In *Well I Never* the dominant theme is class-specific, electrical 'tuition'.[18] Amelia Smith is a pretty, lower-middle-class housewife, overwhelmed by fumes in a tiny, insalubrious flat, and oppressed by a husband who refuses to eat the charred food which she cooks for him by gas. So depressed does Amelia become by her 'failure' as a wife that, when she goes shopping, she literally (and comically) totters and stumbles down the street. But electrical salvation is at hand. She wanders into a showroom and is taken under the wing of an immaculately dressed and articulate woman demonstrator who is then symbolically transported back to the dowdy flat. Amelia is subjected to a crash-course in electrical housekeeping. The language in which she is instructed – 'Here's chicky-wicky [for chicken]. Isn't he cute?' 'Mind those pandies' [for hands] – is modishly upper middle-class. Amelia is berated for her failure to appreciate the simplicity and economy of electrical appliances and told that her commitment to the old 'domestic regime' is tantamount to a betrayal of her marriage and herself. But once she has absorbed the rudiments of electrical house-lore, Amelia is, as we might predict, transformed. Her cooking undergoes miraculous improvement; her relationship with her husband is reinvigorated; and she begins to dress in a style which magnifies rather than masks her natural beauty. A changed woman, she now carries herself with total confidence. She has, the voice of the ubiquitous demonstrator informs us, opted for a new and liberated social role.

Deploying a quite different style, *Proof of the Pudding* discards narrative and fantasy and focuses directly on the mechanics of cooking.[19] Two upper-middle-class women sit in a kitchen and chat about the advantages of grilling rather than frying, the ease with which vegetables can be steamed *al dente*, and the delights of home-baking. At the very end of the film, the 'magic of invisible heat' is symbolised via the superimposition of a pylon over the final frame of an electric cooker.

The linking of technological triumphalism to predominantly 'domestic' themes is a stock feature of EDA films of the later 1930s and the immediate post-war period; and although the publicity element is more overt, and directorial qualities greatly inferior, there are stylistic similarities between this *genre* and the mainstream of Griersonian documentary. Thus *News by Wire* is an ambitious attempt to combine naturalistic images of electrical technology, with vertiginous Hitchcockian angles and perspectives in order to convey the 'mystery' and the ubiquity of the new form of energy.[60] In terms of commentary, triumphalism, scientism and nationalism are heavily stressed. Electricity – and here the servant theme is still detectable – is 'powerful, obedient, clean'. The Grid is a 'British triumph'; a switch is still the 'wizard on the wall'. Both in domestic and industrial terms, electricity is depicted as omnipotent. It is, as the commentatory insists in a section devoted to the benefits of ultraviolet treatment, the 'power that heals'; and here the rhetoric produced by an increasingly sophisticated publicity machine carries social as well as medical implications. The extent to which such productions modified dominant perceptions, or were themselves grounded in existing cultural repertoires, remains obscure. The homilies and short stories of the 1920s had presented and offered the possibility of obtaining electricity within constrained parameters – characters, settings, and, above all, language had all been explicitly and unambiguously upper middle-class. So, equally, cinematic electrical publicity made far-reaching assumptions about the interests and aspirations of its audience. We do well to remember here that it was only in the late 1950s that the patrician mode in British cinema – the failure to acknowledge the existence of an autonomous working-class culture, the assumption that middle-class metropolitan *mores* were *lingua franca* in the Midlands and North, an insistence on consensus rather than conflict – finally began to be eroded.[61]

There was a great deal of preaching to the converted or semi-converted in every type of EDA publicity in the inter-war period. The 'literary' offerings of the 1920s assumed a high degree of stylistic receptiveness on the part of their audiences, with copy-writers working within highly formalised frameworks and often seeming to be as concerned with devising an effective denouement as hammering home a commercial message. In terms of literacy, language, style and illustration, the 'official' pre-photographic image of electricity was beamed directly and unambiguously at the social elite. By the early 1930s, however, the wordy electrical parable or moralistic short story was being replaced by photographic material which gave more immediate visual access to such topics as the precise mode of operation of electrical appliances, the uses of electricity in industry and transport, and how the new energy source was likely to interact with and transform the larger society. But despite marginal concessions to suburbia, settings, structures and, above all, linguistic conventions remained rooted in middle-class experience and cultural expectation. The talkies offered the possibility of developing a style and language which would have made electricity accessible in technical, functional and domestic terms to a truly national constituency. But this was an opportunity which was either lost or ignored. The same stereotypes – the all-knowing middle-class demonstrator, the elegant 'modern' mistress of the house, casually sipping a pre-dinner cocktail, the pipe-smoking professional man – were transposed from the short story and parable of the 1920s to the silver screen of the 1930s.[62]

One means of further exploring the role of the middle-classes as consumers of publicity and mediators of technical knowledge is to focus on a specific and influential inter-war organisation – the Electrical Association for Women – and the life and writings of its dynamic and long-serving director, Caroline Haslett.

## Notes

1  EDA: *Report to the First Annual General Meeting*, 5.
2  *ER*, 21 January 1921, 67.
3  *E*, 14 January 1921, 70.
4  EDA Council *Minutes*, 21 November 1924; R. W. Whitley, 'Electrical Advertising', *ET*, 29 May 1924, 647; and *E*, 18 July 1924, 63.
5  EDA Council *Minutes*, 20 March 1925 and EDA: *Report to the Sixth Annual General Meeting*, 8.
6  *ET*, 12 November 1925, 563.
7  Ibid, 27 October 1927, 514.
8  *E*, 27 January 1928, 82.
9  EDA: *Report to the Eighth Annual General Meeting*, 5.
10  Back cover to EDA: *Report to the Ninth Annual General Meeting*.
11  *E*, 28 March 1930, 385.
12  *EII*, 17 December 1930, 2107.
13  'EDA's New Chief', *E*, 19 December 1930, 771.
14  *E*, 13 February 1931, 236.
15  EDA *Minutes*: 'Report of the General Purposes and Finance Committee on

a Proposal for Reconstitution of the Association', 19 May 1933, 1–4.
16   EDA: *Report to the Fourteenth Annual General Meeting*, 8–9.
17   EDA Council *Minutes*, 15 December 1933 and 18 May 1934.
18   Ibid, 4 October 1933 and 20 April 1934.
19   EDA: *Report to the Fifteenth Annual General Meeting*, 35.
20   'War to the Knife', *E*, 17 November 1933, 605.
21   'Justification by Results', *E*, 30 March 1934, 417. See, also, on this theme: ibid, 27 October 1933, 496 and 3 November 1933, 544.
22   See the discussion in A. C. Cramb, *PIMEA* (1939), 157–8 and the highly critical comments in *E*, 28 February 1936, 282.
23   'The EDA', *E*, 27 March 1936, 397.
24   On this general theme see Andrew Wilson, 'The Strategy of Sales Expansion in the British Electricity Supply Industry between the Wars' in Leslie Hannah (ed.), *Management Strategy and Business Development* (1976), 203–12. For further comment on the 'backwardness' of electrical publicity, see *ER*, 20 March 1923, 482–3 and *Hansard*, 233, 11 December 1929, 528. Herbert Morrison.
25   This material – hand-bills, pamphlets and posters – constitutes a unique source on 'electrical consciousness' in the inter-war period. It is preserved in the Electricity Council Archive.
26   For background on poster art during this period see Bevis Hillier, *Posters* (1969), 224–61 and John Barnicoat, *A Concise History of Posters* (1972), chap.2. Note, also, the excellent collection of reproductions in Forty, *Objects of Desire*, 182–221.
27   'The Letters of a Householder', EDA 23 (1919).
28   'Silent Aids to Comfort', EDA 29 (1919). See, also, 'Just a Turn of the Switch', EDA 30 (1919).
29   This advertisement was produced in the form of a postcard. EDA 143 (1921).
30   'The Home of 1922: Electricity for Ease and Comfort' EDA 222 (1921). On a similar, 'servantless' theme see Susie E. Hammer, 'Electricity and the Life of the People', *E*, 28 September 1923, 330.
31   'House Burned Down = Dinner As Usual, Doris', EDA 209 (1921).
32   'The Man Who Never Missed His Train', EDA 462 (1924).
33   'On Cooks and Cooking: For Health's Sake Use Electricity', EDA 1131 (1933), 12.
34   'Don't Be Old-Fashioned, Mother', EDA 1140 (1933).
35   'From Dirt and Darkness to Sweetness and Light', EDA 72 (1920).
36   'Use the Electric Vacuum Cleaner and You Will Not Fear the Sunshine', EDA 129 (1920).
37   'The Electric Cleaner *beats* the Beater!', EDA 560 (1925).
38   'New Arrivals in the Home', EDA 1132 (1933).
39   'Iron in Comfort the Electric Way', EDA 386 (1923 or 1924). See, also, 'Domestic Power – Why You Will Use It', EDA 442 (1924).
40   'The House You Want', EDA 800 (1929).
41   '3 Plug Points', EDA 905 (n.d., probably 1929).
42   'Electric Cooking', EDA 992 (1932). The theme of 'emancipation' in the home was closely linked to the idea that the wife who invested in electricity would become a better companion and, by implication, a more relaxed sexual partner for

her husband. See, for example, the pseudonymous 'On Monday My Wife Lost Her Looks', *RE-EF*, April 1937, 263 and 'Life's Little Worries', *RE-EF*, April 1938, 21.

43  'Cooking Made Easy: Electricity As An Aid in the Kitchen', EDA 1014 (1933).

44  Probably the best-known visitor in this sphere, and, in terms of books, articles and lectures, the most prolific was Christine Frederick. See, in particular, her *The New Housekeeping: Efficiency Studies in Home Management* (New York, 1913).

45  EDA 1170 (1933).

46  Lydia S. Horton and Dorothy Vaughan, 'Cooking with Confidence: the ABC of Electric Cookery', EDA 1485/2 (1939), 5. See, also, 'Puts the "OK" in Cookery', EDA 1522 (1939). Whether euphoric praise, then or now, of electric cooking, was justified, is a moot point. Very large numbers of 'amateur' cooks and professional chefs remained committed to the visibility and flexibility of the naked flame. And even now, in the age of the automated electric cooker, a majority of the very finest chefs still opt for the 'primitive' gas-ring.

47  'I'm Busy Cooking the Lunch', EDA 1523 (1939).

48  S. M. Watson, '50 F and Under', *EA*, Summer 1937, 263.

49  'Is Your Larder Safe?', EDA 779 (1929), 7.

50  'The Kind of Refrigerator that will Serve You Best', EDA 1320 (1935 or 1936). The backwardness of Britain compared with America in terms of the domestic refrigerator is brought out in *RE-EF*, July 1931, 37–8. An intense concern with 'food safety' is displayed in 'Electric Refrigerators are *Cheapest* to Buy and *Cheapest* to Run', EDA 1379 (1937); 'Protect His Health with Safe Food', EDA 1400 (1937); and 'How to Protect Your Family from Food Danger', EDA 440 (1937). See, also, the engaging though somewhat bizarre 'germ parable' by L. H. Hornsby, 'Immaculate Little Man', *RE-EF*, August 1938, 42–3.

51  'Cold Cookery', EDA 1337 (1936).

52  'It's So easy and Cheap to have ... Electric Hot Water', EDA 1498 (1938 or 1939), 9.

53  Sir William Ray ('and thousands of ... ladies in the land'), 'This Wonderful Age', EDA 1206 (1934).

54  'Such a Lot of Washing Up', EDA 1506 (1939). See also, 'Life is Easier with Electric Hot Water', EDA 1503 (1939).

55  For subtle contextualisation on this and related issues see Ruth Schwartz Cowan, 'Two Washes in the Morning and a Bridge Party at Night', *Women's Studies*, 3, 1976, 147–72; 'The "Industrial Revolution" in the Home: Household Technology and Social Change in the Twentieth Century', *Technology and Culture*, 17, 1976, 1–23; and *More Work for Mother: The Ironies of Household Technology from the Open Hearth to the Microwave* (New York, 1983).

56  On planning and slum clearance see 'Electricity and Housing – Supplement to the *Architect's Journal*', EDA (1934) and 'Rehousing with the Aid of Electricity', EDA 1310 (1935). On office heating systems see 'Heat to Measure', EDA 1299 (1935); and on the medically authenticated prevention of eye-strain, 'Seeing Begins At Home', EDA 1397 (1937).

57  EDA, *Plenty of Time for Play* (1934) (20 min). The film was linked to a pamphlet of the same name, EDA 1247 (1934).

58  EDA, *Well I Never* (1934) (20 min).

59   EDA, *Proof of the Pudding* (1938) (9 min).

60   EDA, *News By Wire* (1938) (17 min). (A selection of short EDA films are on permanent show in the Electricity Gallery of the Greater Manchester Museum of Science and Industry.) On inter-war documentary see, in particular, Stuart Hall, 'John Grierson and the Documentary Film Movement' in James Curran and Vincent Porter (eds), *British Cinema History* (1983), 99–112 and Donald Mitchell, *Britten and Auden in the Thirties: the Year 1936* (1981).

61   For a concise survey of this important phase in the history of British movies see Eric Rhode, *A History of the Cinema from its Origins to 1970* (1976), chap. 15.

62   Similar problems plagued EDA cinematic publicity in the immediate postwar period. In the patriotic and triumphalist *Their Invisible Inheritance* (1945) (20 min, incomplete) actors of cameo roles still pitched their accents uneasily between middle-, lower-middle- and working-class stereotypes. But so also did 'mainstream' performers in Ealing comedies and some of the more 'realistic' films of the early 1960s. Full emancipation from middle-class elocutionary norms in the British cinema is a relatively recent development, closely and subtly linked to ethnic cross-fertilisation. (See and hear, for example, *My Beautiful Laundrette* directed by Stephen Frears from a screenplay by Hanif Kureishi, 1985.)

# 3
# Targeting women

Caroline Haslett, the charismatic director of the Electrical Association for Women from 1925 until 1956, may appear to have powerfully represented the forces of electrical triumphalism and cultural modernity which have been described in earlier chapters. Yet in her writings on the new energy source and, via influential evidence to a wide range of governmental inquiries, Haslett intermittently displayed a preoccupation with issues which transcended the crude rhetoric of electrical progressivism. Her concerns might well have distanced her from a predominantly upper-middle-class membership, who wanted to do little more than dabble in the 'mysteries of electricity'. The 'Director', though, was exceptionally adept at shaping 'electrical women' to her own ends. She was also deeply attracted to an emancipated independence which could only be fully experienced by the moneyed élite. Her formidable technical and scientific knowledge, administrative skill and vast energy enabled her to exercise effortless authority over disciples who were frequently her social superiors. This personal dominance also ensured easy access to the upper echelons of British metropolitan society in the 1920s and 1930s. And it was precisely this 'social success' which blunted the cutting edge of Haslett's critique of explictly sociological aspects of the 'electrical revolution'.

By the time she was 35 this Sussex-born daughter of a railway signal fitter, and pioneering activist in the co-operative movement, had become part-founder and director of the first national organisation to express the 'woman's view' on electricity. Concentrating on clerical subjects at school, she worked in that capacity in her first job with the Cochran Boiler Company. But the emancipatory climate created by the social and occupational upheavals of the First World War proved to be a lasting one and a speculative application led to her appointment as the first secretary of the Women's Engineering Society, which had been 'formed [in 1919] ... to conserve for women the right then gained of serving the nation and the community in the way best possible to the individual'.[1] In 1924 a member offered a paper to the WES on the domestic use of electricity and the possible formation of an organisation to express the 'woman's view' of the new form of energy. The Society agreed to investigate the possibility of establishing such a body and set up a subcommittee under Haslett's chairmanship. There followed a meeting in London in November 1924 which was attended by representatives, both male and female, from science, electricity, education and interested voluntary organisations.[2]

A manifesto was issued to the press early in 1925. 'Man's most promising adaptation of Nature's bounty', it read, 'becomes each year a more potent force in every sphere of work. Its possibilities as a saver of labour and aid to fuller life for the home make an appeal to all women, and particularly to the increasing number who take an interest in public affairs.'[3] Caroline Haslett had dominated the WES subcommittee, and by 1925 had taken up her place as director. A provisional constitution was drafted; and, by the end of the following year, branches had been established in Glasgow, Birmingham and Manchester and Salford.[4] But finance proved to be a major problem – and stories of the director travelling to far-flung branches and paying her expenses out of her own pocket were already contributing to a burgeoning cult of the Haslett personality. It was fortunate, in this respect, that one of the representatives at the inaugural meeting had been J. W. Beauchamp, founder and director of the EDA; by the beginning of 1926, his organisation, cannily realising that women would play a crucial role in the diffusion of knowledge about domestic electricity, was subsidising the EAW to the tune of £400 a year.[5] Although Haslett still felt it necessary in 1927 to warn her membership that 'we are still badly handicapped for funds, and ... would urge every member to help the good work which we are doing', her own position had become more secure.[6] Increasing numbers of activities and campaigns were getting under way – a questionnaire had been devised to investigate the electric cooker from 'the Woman User's point of view', and the Association had initiated what would later become the EDA's own highly successful Outlet Campaign.[7]

By the beginning of 1928 the Association was being described as 'this lusty young organization',[8] and, half-way through 1929, sixteen branches were in existence.[9] The organisation, as the technical press rapidly came to realise, could play a crucial role in the dissemination of the 'electrical idea'. 'Its members', the *Electrical Review* reminded its readers in 1929, 'are peculiarly well fitted for [their] task, as they understand the mentality of the housewife and can enter into her difficulties better than any male instructor ... To support the EAW, therefore, should be a cardinal principle with electrical men.'[10] In 1931 the still exceptionally youthful Haslett was awarded the CBE. 'With Electricity as her Excalibur', enthused one of her disciples, 'she has already worked wonders, and it is in recognition of this unflagging devotion to an ideal that an all-wise Monarch has bestowed his honour'.[11]

'We can claim', Haslett wrote in the *Annual Report* for 1932, 'to have built up a *"women's point of view"* on matters electrical which is undoubtedly influencing the more rapid development of domestic electrification in this country'.[12] National and international conferences were now more regularly and systematically held; an electrical housecraft diploma had been established; and, early in 1933, the EDA increased its grant to £1450 a year, and donated a capitation payment of a shilling for every new member.[13] In her *Annual Report* for that year Haslett announced in classically triumphalist style, that thirty-one branches had been established and that 'the Association is filling a very necessary niche in our Modern Life by providing a field for the expression of women's interest in the applications of this greatest of all modern sciences – Electricity'.[14] The London headquarters was moved from the existing cramped premises in Kensington, originally secured for the Association by the veteran electrical pioneer, Colonel R. E. Crompton, to a much larger building in Regent Street.[15] There is a contemporary photograph of the new club room. It shows groups of exceptionally well-dressed women languidly lounging in skeletal chairs, and browsing through magazines by low tables decorated with vases of fresh flowers. The high ceiling and slimly elegant side-pillars lend an impression of breadth and height and direct the observer's eye forward and upwards, first to a magnificently cabineted art deco radio, and then – a culminating touch – to a formal portrait of the director herself.[16]

In May 1934 there were rumours, for the first and only time, of internal dissension. But the electrical press sprang to Haslett's defence and pointed out that a 'certain amount of ... criticism' was a legitimate 'means of preventing control becoming vested in the hands of an oligarchy and of keeping those at the head in touch with the view of the rank and file'. Such things were evidently 'fully realized by the present leaders of the EAW'.[17] Undeterred, the central administration ventured into new territory, setting up an enquiry bureau and sponsoring a questionnaire-based report into the design and performance of electric appliances.[18] Only now, in 1935, a year in which no fewer than nineteen new branches were set up, did the Association commit itself to an investigation of the provision of electricity for the urban and rural working class; the resulting publication, *Report on Electricity in Working Class Homes* by Elsie Elmitt Edwards soon established itself as a major social and quantitative statement.[19] In the following year, Haslett, as if to correct the impression that both the metropolitan centre and the branches spent too much time looking round power stations and chocolate factories, gave penetrating private evidence to the McGowan Committee on the distribution of electricity. 'The fact of the existence of our Association', she insisted, 'is a criticism of the electrical industry.' She went on to savage lazy local companies which deprived householders of a cheap and effective domestic supply, and appliance manufacturers who paid scant regard to the needs of consumers.[20]

At branch level – and there were now more than sixty of them – 'the romance of electricity' was still casting its spell 'from generation in the stations that so merit their description of power houses, through all the stages of distribution, particularly in that phase familiarly known as the Grid, to the final adaptations magically controlled by the switch'.[21] Scientific management in the household was now also making itself felt. A lavish 'all-electric' demonstration house in Bristol, partly funded by one of the Association's many benefactors, introduced into 'the routine of housework the results of scientific study that have already revolutionised industry in factories' and 'brought within the bounds of possibility that leisure which is repeatedly promised to this generation but which has hitherto proved illusory'.[22] It was all the more ironic, therefore, though deeply revealing of the EAW's fundamental class and ideological orientation, that attention was now turned to the electrical education of domestic servants, or 'home workers' as they had been redesignated.[23]

There were clear connections between the 'home workers' campaign and the concern expressed a year or so earlier *vis-à-vis* the provision of domestic electricity for the working class. The EAW certificate, which was awarded to servants who showed themselves able to master washing-machines, vacuum cleaners and percolators, may have contributed in the eyes of most members to the 'elevation' of domestic labour. But the linking of this initiative to the 'servant problem' revealed that the Association was still overwhelmingly wedded to élitism and social conservatism.[24]

Caroline Haslett and others claimed from the very beginning that EAW members were drawn from 'all shades of opinion and . . . all ranks in life'.²⁵ 'We have many poor women among our members', Lady Moir, the Association's third president insisted in the late 1920s, 'and I think of all the work these women have to face on the grey days. Electric washing machines would be an inestimable boon to them, but they are too expensive.'²⁶ *The Electrician* surely came much closer to the truth when it stated in 1934 that 'to a large extent it [the EAW]has found it necessary to make itself an officers' training corps and in doing so it has necessarily been in some danger of losing touch with the lower rungs of the social ladder'. The organisation should 'avoid all suspicion of patronage like the plague' and ensure that its material 'is not written in a manner that is entirely incomprehensible or even irritating to a large proportion of those for whom it is intended'.²⁷ But the tone had already been firmly set by the powerful aristocratic and upper-middle-class influence which Haslett herself had done little to discourage. 'Women appear to have electrical homes for different reasons', the Association's first biographer, Peggy Scott, wrote in 1934.

The Dowager Lady Swaythling rejoices that 'there is not a cold corner of her house' since she brought electricity into it. Lady Mount Temple (the former Mrs Wilfred Ashley) delights in the beauty of light in the right place. Lady Belhaven and Stenton congratulates herself that she has revolutionised the labour of a big country house. Mrs Herbert Morrison is glad of electricity because it gives her time for public work.²⁸

Lady Mount Temple had put electricity to extraordinarily sophisticated and elaborate decorative uses. Carefully placed lights, Scott noted, caused

lovely reflections in the high-recessed panel of grey mirror glass which is the centre of decoration [in this room] with its silvery white walls. Sitting beneath the panel on a black glass shelf is a 'Blanc de Chine' god, his broad grin enhanced by a single light placed beneath him. Silver reindeer and silver swans are on the descending shelves, and the electric fire is placed behind glittering leaves, chromium plated. More light comes from a large piece of uncut green glass on a black glass table. A jade lamp on the gleaming black and mirror glass writing table has a daylight bulb so that the white background of the room is not given a yellow shade.²⁹

Within this social élite, electricity was applied to the widest possible range of aesthetic, 'beautifying' and romantic purposes. 'On one occasion', another EAW member, Mrs C. C. Paterson, recalled,

when giving a children's party in January, the French windows were opened for a few moments to clear the air of the fumes of indoor fireworks, and before anyone could stop them, the children in their silk stockings and slippers were streaming over the floodlighted lawn, and making for a distant shadowy wood. One of the parents, seeing my distress, said: 'It only needs the fairies to make the scene perfect.'[30]

Women such as these – public figures, professionals, socialites – also exploited the full cosmetic potential of the new source of energy. 'Does her complexion need a tonic?', Violet Desmore asked a hypothetical fellow-EAW enthusiast in 1930, 'What could be better than the electric vibrator? Is her figure a little too plump here and there? Well, the electric vibrator belt is simple and effective. By means of yet another electrical device she can give herself a sunbath at any time of the year. With the aid of an electric dryer she can finish off that hair-shampoo in double quick time.'[31] Appliances should be used in every part of the home and for every conceivable function – electricity was 'modern' and fashionable and it yet further enriched already highly privileged and affluent life-styles. 'Given an electric kettle', Elizabeth Craig wrote in 1935, 'toaster, coffee percolater, waffle iron and a table stove, and a dinner or tea wagon, entertaining with or without the help of a maid should be ABC.'[32] Electricity, in a social environment such as this, became a quasi-magical elixir which would abolish not only housework but every discomfort and inconvenience which detracted from leisure and narcissism. In 'The Metamorphosis of a Housewife' the popular American journalist and lecturer, Elizabeth Sloan Chesser described an imaginary tableau in which 'a housewife [read] a novel against a background of sweated women carrying buckets, mangling and ironing in the bad old way. She was shown again with a tennis racket saying *"au revoir"* to her joint and fruit pies cooking themselves obediently for the husband in his office, her children coming from school, her mother-in-law on her way to join the family repast.'[33]

Among the aristocracy and the upper middle class, electricity could also be deployed to improve the quality of family relationships. 'Small jobs of [electric] nursery cooking', wrote a contributor to the *Electrical Age* in 1937, 'make nurse independent of cook for the smooth running of the home, or save the single-handed mother from journeys up and down stairs and the use of the larger and more costly appliances'. 'The mother or nurse of to-day', she went on, 'is a happy, healthy person who has health and energy to be really interested in mothercraft and appreciate what a fascinating business it is when half the fatigue and boredom of routine is lifted in this way from her shoulders'.[34] In this part-real, part-idealised, part-absurd all-electric cosmos, every mundane task would eventually become extinct. 'And what about the EAW?', another contributor to *The Electrical Age* reflected in 1938, 'How do we wash up?

We don't . . . [Each member has] a washing-up machine . . . You should just see it work while we stand idly by, our lovely hands fresh from the manicurist, unsoiled, not even wet from start to finish!'[35] 'It is evident', another member wrote in euphoric mood, 'that over a few weeks hundreds of hours of drudgery are turned to hundreds of hours of leisure [by the application of electricity]'.[36]

By contrast, the Association's depiction of the electrically un-enlightened and socially unsophisticated could often be noxious: upper-middle-class members regularly patronised and caricatured the people they came into contact with on their 'electrical travels'. 'A very human touch', wrote Cecile Francis-Lewis after an EAW visit in 1929, 'was the welcome given to us by the villagers. They had (entirely on their own initiative) white-washed their cottages and hung up . . . bunting. As I passed a very old women touched my hand, and said: "Ay, but we are glad to see yer all; good luck to yer". I looked up to thank her and saw written on a piece of brown paper fastened to the Union Jack: "Welcome to the Glasswork Visitors" '.[37] In the first published account in the Association's journal in 1934 of working-class attitudes towards electricity, a Lancastrian was made to express himself in a bizarre mixture of regional cliché and standard English: 'By gum, but it's champion . . . Eh, it's been grand, hasn't it lass? During the hot weather we could have sufficient hot water for baths in just over an hour, without the discomfort of a coal fire.'[38] The same author wrote that 'it is typical of these good natured Lancashire folk to give a helping hand when occasion arises'.[39] And an article entitled 'The Working Woman Speaks on Electricity' was again characterised by a hodge-podge of flawed (cockney?) pseudo-regionalism and ghosted reportage. 'What I says is . . . why shouldn't the working classes have some of the luxuries that the upper classes have, and specially the advantage of electricity in their homes.'[40] As if this were not enough, hyper-enthusiastic, pro-electric attitudes were inserted into allegedly factual accounts of working-class experience: 'Thanks to the advent of the Electric Washer, a new Era has opened out, and washing day is a "thing of Beauty and a Joy for ever".'[41] All this belonged more to the club room in Regent Street than a Lancastrian back-to-back.

But it was servants who fared worst of all. Margaret Partridge, an accomplished engineer and journalist, and influential member of the Association in its earlier days, expressed the professional-cum-bachelor girl's unthinking arrogance towards domestic help. 'It is strange', she wrote in 1927, 'how hard it is to make Mary Jane use a vacuum cleaner intelligently. Certainly she must have had Mohommedan ancestry, as her passion for going down on her knees knows no bounds.'⁴² De Ferranti's widow, Gertrude, believed in the efficacy of 'breaking in' servants who had returned from 'pleasanter jobs in the factories' during World War I. 'I find', she wrote in 1928, 'that once they have used electricity for cooking, the workers never want to return to coal, and that the maids are not as wasteful in using it as they were in using coal.'⁴³ Caroline Haslett's 'char', if we are to believe the account contained in the first history of the association, was a paragon of electrical virtue: she had evidently stopped kneeling and made a rapid adjustment to the vacuum cleaner. 'What I like about these electrical things', she is supposed to have said, 'is that it makes you self-respectin'! You can put your clean apron on and keep it clean, and you don't have to go on all fours, like some monkey.'⁴⁴ The EAW's intention, more insistently emphasised after the launch of the Home Worker's Campaign in 1936, was to produce 'an intelligent and interested rather than a grudging worker' and to raise the status of domestic labour.⁴⁵ But the social background of the metropolitan élite, which played a crucial role in shaping policy, worked against any radical reformulation of the relationship between employer and employee. Vacuum cleaner or no vacuum cleaner, servants in inter-war Britain were more often defined as part of a 'problem' than in terms of their human needs.

'Electricity and all the natural elements', wrote another EAW enthusiast in 1938, 'give life, colour and beauty. Never think of electricity as artificial light that we can make come forward when the sun is absent. Electricity permeates all life and has been stored and harnessed by man for his joy and use.'⁴⁶ This quasi-hymnal tribute to the new form of energy typified the explicitly romantic-cum-scientific vision of electricity adhered to by numerous supporters of the Electrical Association for Women in the inter-war years. Whether welcoming foreign 'electrical enthusiasts', visiting Battersea Power Station, or listening to lectures on improving the figure via the use of vibrator belts, members were committed to an ideology which involved the conquest of technical ignorance, the modernisation of the home and the enrichment of leisure. Electricity was both 'servant' and elixir. Looking back from the vantage-point of 1939, Caroline Haslett identified the EAWs constituency as the 'housewife and the professional woman, the townswoman and the countrywoman, the bachelor girl and the mother of a large family, in fact ... all women'.⁴⁷ 'Gradually', she wrote in equally sanguine mood in the same year, 'has built up an organisation whose thousands of members equip themselves to radiate electrical knowledge'.⁴⁸ The precise nature of that 'knowledge' had itself been deeply influenced by Haslett's own views on the relationship between the new energy source and broader cultural change. In the rough draft of an essay published in 1938 in a collection of collaborative essays entitled *Where Do We Go from Here?*, she ruminated on the 'achievements' and the potential of the new source of energy. 'Electricity', she wrote, 'is the great civilising force, working not in a violent revolutionary fashion, but by the powerful permeation firstly of the minds, then the methods and finally the homes of the people.'[49] Moving on to the connections between electricity and women's emancipation, Haslett claimed that the First World War had played a central role in opening up opportunities in the professions. But then, in the 1920s, there had been a reactionary closing of the ranks, and without the liberating potential provided by such organisations as the WES and the EAW, the cause might well have suffered a grievous setback.[50] 'Where do we go from here?' marked the rhetorical conclusion of the essay and the response conveyed the full flavour of the scientistically teleological romanticism which was central to EAW ideology. 'For each one of us Utopia is different, but if we proceed electrically we are assured of safe arrival and pleasant travelling.'[51]

Allied to the triumphalist, emancipatory and educational strands in Haslett's thinking was a strong interest in the application of scientific management to housework. The growth of mass production, Taylorism and time-and-motion study in the United States and Britain since the end of the first World War had convinced advocates of 'scientific household management' that there were few fundamental differences between factory work and work in the home. It was argued, further, that the adoption of such techniques had been inhibited by the fact that women, rather than men, had habitually borne by far the heavier domestic burden. The underdeveloped state of scientifically planned household labour was, in this sense, a reflection of the maldistribution of social power between the sexes. But it would be misleading to link Haslett's ideas too closely to the late-twentieth-century feminist ideologies on the nature and status of housework. The earlier 'scientific' ethos placed much greater emphasis than the later and more politically radical movement on the extent to which domestic technology could be expected, first, to transform and, finally, to abolish domestic labour. Ideologically, emancipatory feminism was outweighed by scientism and technological triumphalism.[52]

'Household work', Haslett wrote in 1931, 'should be performed on just the same scientific basis as workshop operations, and the same amount of care should be given to the household worker as is now bestowed upon that of the workshop operative'.[13] But there would be little progress without a rationalisation of the 'domestic' laboratory which lay at the heart of every household – the kitchen. Here, Haslett insisted, there was 'great scope for the service of electricity . . . electrical authorities should take an interest, not only in selling electrical energy, but in kitchen planning, which should imply *electrical* kitchen planning'.[14] Within the modern, electrified and rationalised kitchen, then, the housewife would first learn and then dedicate herself to scientific movement. Rhythm was all-important – 'hence marching songs, sailors singing when hauling the anchor, and the rhythmic action of the skilled wood-cutter'. So, also, was economy of movement, with as few members of the body as possible being brought into action during any single task. It was essential to maintain a balanced posture and to work with both hands, rather than relying exclusively on left or right. Due attention must be given, in this respect, to 'anatomical structure' and to the fact that every finger, and not just the 'easy' ones, should be used during lengthy operations. Materials and tools must always be within easy reach and every advantage 'should be taken of the human tendency to form habits'.[15] If each of these principles were adhered to, the kitchen would become 'a workshop as well as a centre of the work of the house'.[16]

In her post-war articles and books, and notably in the eccentrically titled *Problems Have no Sex*, Haslett re-emphasised the importance of economy of movement in housework and linked these ideas to the need for psychological liberation and national efficiency and fitness. 'Electricity', she wrote in 1947, 'is a force, a form of energy which can be used to replace human energy in a multitude of household tasks', adding, in an uncharacteristically direct reference to the sexual politics of housework, 'it has been said with justification . . . that old-fashioned non-labour-saving methods of work in the home would have vanished long ago if the husband and not the housewife had to work in the home'.[17] There continued to be 'much scope for education of the housewife in planning her housework along labour-saving lines; and in the teaching of the principles of motion study in the home so that the maximum benefit can be gained from the use of equipment provided'.[18] But domestic liberation, via scientific movement and the methodical organisation of the household, would only be attained via thoroughgoing psychological reorientation. 'Women must . . . emancipate themselves from the idea that the expenditure of a vast amount of energy on housework is in itself a virtue . . . All this labour occupies hours of time and absorbs the energy of a woman leaving her unduly drained of vitality and unable to give of her best to the important components of the home – her husband and her children.'⁵⁹ This flow (and waste of energy) within the individual home must also be comprehended in national terms. 'All forms of energy', Haslett asserted in *Problems Have no Sex*, 'are national assets and human energy is the most precious because there are definite limits to it: it is expendable.'⁶⁰

In this, one of her final extended works, Haslett attempted to set the 'domestic problem', which was still, in her view, largely unresolved, in full historical context. Drawing on Olive Schreiner's model of the exclusion of women from an ever wider range of occupational opportunities, Haslett held that, in modern times, 'instead of men's work and women's work developing side by side in hemisphere as it were, the tendency at any rate since the industrial revolution has been for it to develop in strata, with women as operatives and subordinates under men administrators and directors'.⁶¹ But she still held back from an explicitly ideological and feminist analysis and returned, as one might expect, to her well-tried scientific and utopian mode. 'Just as the introduction of machines temporarily deprived woman of her more skilled status as a producer, so the general advance of science adapted to the needs of mankind may enable her not so much to regain a lost status, but to create for herself a new and important place in society.'⁶² Two years later, in 1951, Haslett was still insisting that electricity was 'something more than a contribution towards the emancipation of women. It is a factor in race survival itself.'⁶³ Noting that social stability was now 'sadly lacking', she argued that large-scale problems could only be solved 'by the enlightened application of the genius of the scientist and the engineer to meet the needs which the sociologist lays bare'.⁶⁴ Haslett was now convinced that she could discern a severe 'national deterioration'. If there really were a lasting solution, it was indistinguishable from the generalised panacea that she had been championing, in one or another form, for nearly thirty years. 'In considering the nation's needs it is not a question of whether we can afford to have electricity, but a realization that we cannot afford to do without it if we are to have any social progress at all.'⁶⁵ Here, as before, concrete social analysis was outweighed by triumphalism.

## Notes

1   *E*, 29 September 1939, 300.
2   Wilfred Randell, *Electricity and Women: 21 Years of Progress* (1946), 17–23. See, also, for coverage of the same events, *RE-EF*, April 1929, 346.
3   *ET*, 29 January 1925, 126.

4   Randell, 59.
5   EDA Council *Minutes*, 16 October 1925; EDA: *Report to the Fifth Annual General Meeting*, 14; and EAW: *First Annual Report*, 2.
6   EAW: *Second Annual Report*, 6; EDA Council *Minutes*, 15 February 1929; and EAW: *Fourth Annual Report*, 4.
7   EAW: *Fourth Annual Report*, 4.
8   *E*, 24 February 1928, 204.
9   EAW: *Fourth Annual Report*, 4.
10  *ER*, 1 February 1929, 181. For a 'pre-EAW' statement of this type see *E*, 7 September 1923, 237–8.
11  Gladys F. Sharp, *EAW*, April 1931, 143; Randell, 80.
12  EAW: *Sixth Annual Report*, 44.
13  EDA Council *Minutes*, 20 January 1933.
14  EAW: *Eighth Annual Report*, 9.
15  Rosalind Messenger, *The Doors of Opportunity: a Biography of Dame Caroline Haslett* (1967), 65; and EAW: *Eighth Annual Report*, 9.
16  *Contribution to Progress: an Account of the Aims and Activities of the Electrical Association for Women* (1937).
17  'The EAW Conference', *E*, 12 May 1933, 609–10; and Randell, 59.
18  EAW: *Ninth Annual Report*, 4; *The Design and Performance of Domestic Electrical Appliances* (1934).
19  Elsie Elmitt Edwards, *Report on Electricity in Working Class Homes* (1935).
20  Private evidence by Caroline Haslett to Ministry of Transport: *Report of the Committee on Electricity Distribution* (HMSO, 1936). The transcripts referred to here are to be found in the IEE Archive, NAEST, 93/6.1.
21  EAW: *Eleventh Annual Report*, 10.
22  Ibid and *Twelfth Annual Conference of the Electrical Association for Women: Practical Aspects of Kitchen Planning* (1937).
23  *EA*, July 1936, 89.
24  EAW: *Thirteenth Annual Report*, 6.
25  EAW: *Third Annual Report*, 1.
26  Peggy Scott, *An Electrical Adventure* (n.d., probably 1934), 35.
27  'Downward Penetration', *E*, 27 April 1934, 552
28  Scott, 73.
29  Ibid, 74. For a similar approach to 'electrical interior design' see Mrs Wilfred Ashley, 'Presidential Address', *Report of Proceedings of the Fourth Annual Conference* (1929), 3.
30  Scott, 68.
31  Violet Desmore, 'Tell Us More About Electricity', *EAW*, October 1930, 71.
32  Elizabeth Craig, 'Entertaining in the Spring', *EA*, April 1935, 827.
33  Elizabeth Sloan Chesser, 'The Metamorphosis of the Housewife', *EA*, October 1935, 918.
34  Len Chaloner, 'Childhood Today', *EA*, January 1937, 182.
35  Dorothy Brooke, 'The Secret Slum and How to Banish It', *EA*, Autumn 1938, 478.
36  Hedwig Auspitz, 'More Leisure for the Housewife', *EA*, Spring 1939, 553.

37  Cecile Francis-Lewis, 'Where A Fairy Tale Came True', *EAW*, October 1929, 542.
38  Marjorie M. Hutchinson, 'What the Lancashire Working Woman Thinks of Electricity in the Home', *EA*, January 1934, 600–1.
39  Marjorie M. Hutchinson, 'The Electric Cottage Offers Leisure', *EA*, April 1934, 645.
40  Muriel Watson, 'The Working Woman Speaks on Electricity', *EA*, July 1934, 705.
41  Marjorie M. Hutchinson, 'The Modern Wash-Day in a Working Home', *EA*, April 1935, 851.
42  M. Partridge, 'Electrical Spring Cleaning from the Woman's Viewpoint', *ET*, 31 March 1927, 469. For further scathing comment on the 'non-electric' servant see *E*, 12 June 1931, 837–8. Less dogmatic comment may be found in *ER*, 25 June 1926, 943.
43  Gertrude Z. Ferranti, 'Electricity in the Household', *ER*, 19 October 1928, 639. See, also, in very similar vein, 'Why Domestic Servants Are Scarce: A War-Time Taste of Freedom', *Manchester Guardian*, 6 January 1921.
44  Scott, 38.
45  *EA*, July 1936, 100.
46  Mrs Watson Ingram, 'Beauty', *EA*, Autumn 1938, 475.
47  Caroline Haslett, 'Electrical Association for Women', *ER*, 6 June 1939, 787.
48  Caroline Haslett, *Household Electricity* (1939), ix.
49  Caroline Haslett, Draft contribution to J. Mayers and B. Spiers (eds), *Where Do We Go from Here?* (1938). IEE Archive. NAEST 33/15.2,1.
50  Ibid, 4–5.
51  Ibid, 17.
52  There is very little scholarly historical writing on this important subject by British writers. American experience has been fully and impressively explored by Dolores Hayden in *The Grand Domestic Revolution: a History of Feminist Designs for American Homes, Neighborhoods and Cities* (Cambridge, Mass., 1981) and Ruth Schwartz Cowan, *More Work for Mother*, chaps 4–6.
53  Caroline Haslett, 'Electricity in the Household', *Journal of the Institution of Electrical Engineers*, 69, 1931, 1377.
54  Caroline Haslett, 'Economic Domestic Electricity', *ER*, 15 January 1932, 87.
55  Messenger, 80. See also the bizarrely stereotyped 'kitchen movements' in the EAW film *T'was on a Monday Morning*, (1945) (20 min.) (Electricity Council Archive).
56  Caroline Haslett, *ER*, 6 August 1937, 189.
57  Caroline Haslett, 'Electricity in the Home', *Journal of the Royal Society of Arts*, 95, 1947, 652.
58  Caroline Haslett, *Problems Have no Sex* (1949), 43.
59  Ibid, 41.
60  Ibid, 44.
61  Ibid, 61.
62  Ibid, 62.
63  Caroline Haslett, 'Electricity – A Factor in Social Progress'. British Electrical Power Convention, June 1951. IEE Archive. NAEST 93/6.2, 20.
64  Ibid, 21.
65  Ibid.

# 4

# Urban experiences

Urban and 'old' suburban Britain electrified much less rapidly than triumphalists or constructors of positive images of the new source of energy predicted. 'Probably 80 to 90 per cent of the homes in the [typical] area are unwired', lamented the *Electrical Times* in 1922, 'of those that are wired, the majority use nothing but light'.[1] A year earlier Margaret Partridge, an astute observer of the electrical scene, asked herself why so many densely populated areas were still deprived of a full supply, and emphasised the existence of a sophisticated gas infrastructure, fuelled by strategically placed coal-mines, 'public ignorance of all technical details of electricity', and excessive cost.[2] There were, of course, exceptions to the complaint that the supply industry was lagging badly behind its European rivals: positive achievements, such as the creation at Welwyn Garden City, of an 'all-electric' town, received wide publicity.[3] But urban and suburban Essex was demonstrably undersupplied in the later years of the decade; much still remained to be done.[4] The technical press exhorted companies and contractors to cut costs and devise schemes which would encourage larger numbers of 'average' consumers to have their houses wired. The industry had to become more sales-conscious and respond positively to every inquiry – even from 'electrically ignorant' and barely coherent correspondents who wanted to know 'what is it going to cost me to have a Metor in for heating the Bedroom I have been told that you put Metor in free if it is going to cost me What it did for Lighting it will have to go by as I can't afford it . . .'.[5]

By 1931 the *Electrical Times* was reporting that 'of 10½ million houses in this country, less than 30 per cent are wired, and not 1 in 1000 is all-electric. There are nearly 8 million houses to go for, most of which are either within reach of distributing mains or in areas in which main-laying would prove profitable.'[6] As for householders who were connected up, three-quarters of them had, according to the same source, 'never been properly canvassed and followed up: the consequence is that their average yearly consumption of units is measured in hundreds instead of thousands, as it easily could be. As for the non-users: what a gold-mine awaits some honest spade-work.'[7] Regional patterns of consumption and application had in many instances been established in the pre-war era – the load in Northern towns, for example, tended to be more heavily skewed towards industrial and commercial functions than in the South. But the scanty provision of domestic electricity in a depressed community like Leigh as late as 1934 could only be interpreted as a condemnation of the sales and pricing policies of the local concern. The total population was 45,000, and they lived in 11,000 households; but no more than 4700 of these were supplied with any form of supply.[8] Here, as in many other parts of the North-West, the North-East, South Wales and Scotland, the new source of energy tended to be seen as a rich man's commodity, overpriced, culturally alien, and hedged round by impenetrable supply regulations; these were the factors which dictated that in many industrial regions the working class remained loyal to gas and sceptical towards electricity until the 1940s.[9]

Although she was a fully paid-up triumphalist, the EAW activist, Elsie Elmitt Edwards understood better than most the reasons which accounted for the gulf between modernistic iconographies of the new form of energy and its actual availability to working-class consumers. 'Unless', she warned in 1935, 'builders, landlords, housing committees etc., can be made to think electrically, and put in wiring and certain pieces of apparatus when buildings are being erected, most working class families have about as much chance of possessing electrical apparatus as they have of owning the Mint or the British Museum.'[10] A year later, in a confidential report, which cast doubt on the continuing viability of an industry still divided between the demands of private profit and national electrical development, the McGowan Committee pointed out that 'at the end of 1934, in urban areas alone, there were some 8000 miles of streets in which distributing mains had not then been laid, representing approximately one-fifth of the total mileage of streets in such areas'.[11]

To test this pessimistic assessment against the full range of electrical experience in urban Britain between 1919 and 1939 is not possible. What will be attempted instead is an analysis of a single concern, Manchester, which 'even in 1903 . . . was by far the largest and most important of its kind in this country' and which, in the early 1920s, was still maintaining 'its position at the head of the municipal undertakings'.[12] A description of the social ecology of domestic electricity supply in this large and economically diverse conurbation will allow comparisons to be made with other urban areas. Detailed attention will be given to the highly uneven social distribution of the new source of energy, the role which the Electricity Department played in shaping local attitudes towards domestic supply, the continuing conflict with gas, and the extent to which debates about electricity mediated and clarified larger political issues.[13]

Differences in the 1880s over the balance to be struck between municipal and private enterprise inhibited rapid development of the new utility. In 1890 the Manchester Electric Lighting Order was finally confirmed and plans prepared for the electrification of the city centre. By 1893 a five-wire scheme was successfully operating in a modified form. The 'governing' body in these early days was a subcommittee of the municipal Gas Committee but, in 1897, the Electricity Committee attained independent status and by the turn of the century, the new source of energy was powering the tramway system. Despite labour problems and a highly embarrassing corruption scandal in 1900, there had been a rapid growth in suburban supply by 1914. Insisting that it required every available resource to establish itself as a valid competitor to gas, the Committee successfully fought off demands that it must pay a fixed annual sum out of profits to reduce or stabilise the general rate – the largest amount subscribed in any single year was £12,000. The outbreak of war disrupted further schemes for the expansion of domestic and civil industrial supply but the increased output required at the inner-city Stuart Street Station for munitions production laid the basis for increased capacity in the immediate post-war period.[14]

For several years after the war, ambitious schemes for large-scale domestic electrification were stymied by general economic instability, shortages of raw materials and labour unrest; and, as Figure 4 indicates, it was only in 1923 that the upward trend in aggregate sales was re-established, to be sustained until 1930. Significantly reduced rates of growth between 1930 and 1934 are an indication of the severity of the depression in the urban North-West. But from 1936 until the outbreak of world War II, demand re-established itself on a new and higher plateau – this, as we shall see, was a period in which the fruits of domestic electrification began to be more evenly spread than they had been during the minor boom of the later 1920s. What the data in Figure 4 does not show, however, is the proportion of the total population which had access – in the form of light, heat and power for appliances – to electricity

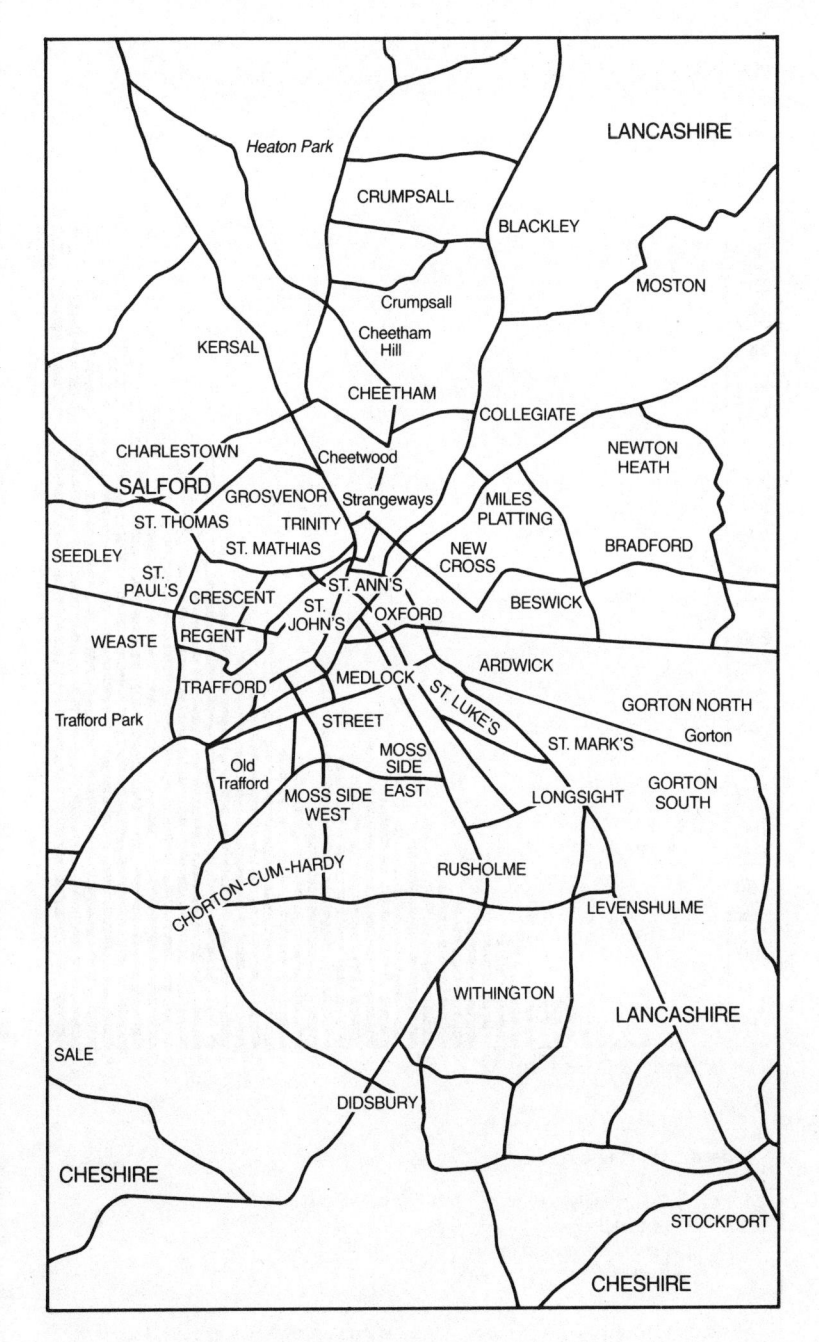

**Figure 3** Manchester in the inter-war years

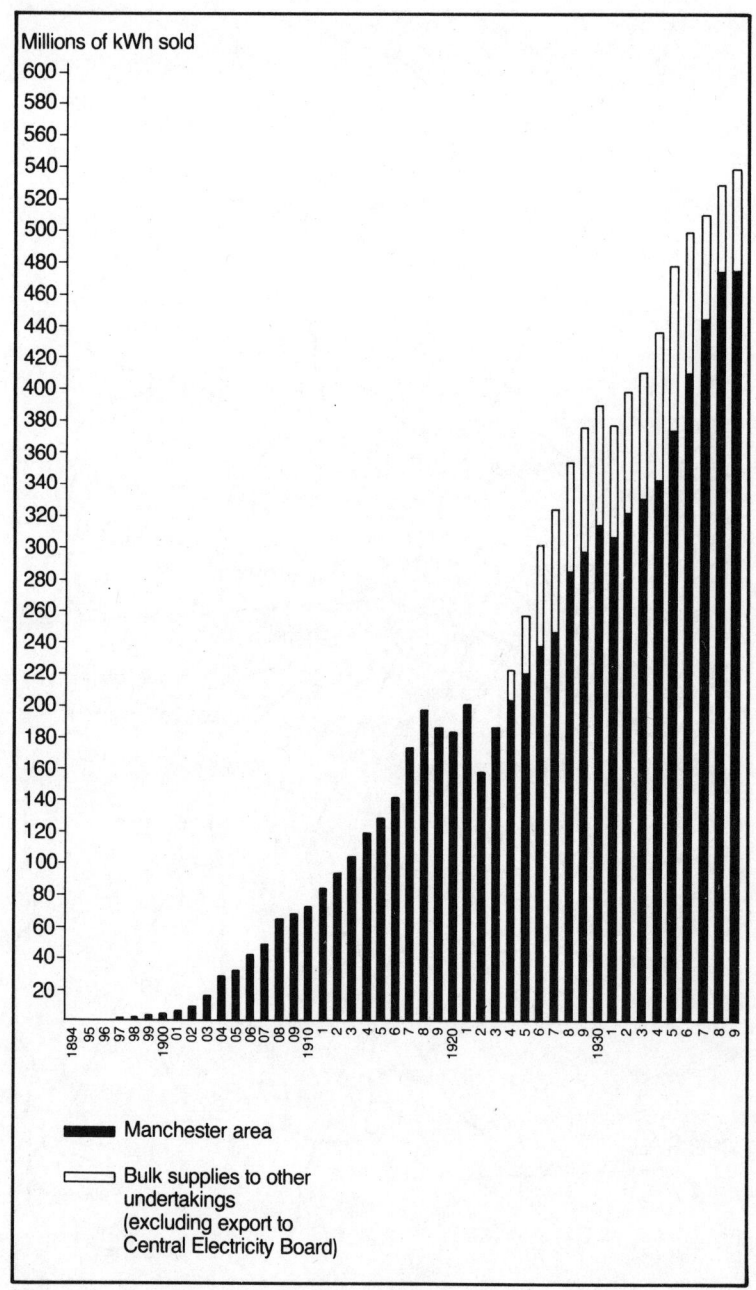

**Figure 4** Growth of electricity sales in Manchester. Source: *How Manchester Is Managed* (1939), p. 76

in Manchester between the eve of the First World War and the later 1930s. The material in Table 2 seeks to rectify this omission.

**Table 2** *'Real' Availability of Domestic Electricity in Manchester 1914–1937*

| Date | Population | 'Official' consumers | Availability |
|---|---|---|---|
| 1914 | 714,000 | 11,200 | 50,400 |
| 1924 | 750,000 | 24,800 | 111,600 |
| 1934 | 770,000 | 80,000 | 363,000 |
| 1937 | 770,000 | 120,000 | 540,000 |

Sources: *Housework Made Easy* (Manchester Electricity Department, 1938); Shena D. Simon, *A Century of City Government; Manchester 1838–1938* (1938), 440; and decennial *Census Reports*.

It reveals that in 1914 approximately one in fifteen of the inhabitants – men, women and children – of the administrative area supplied by the Manchester Electricity Department lived in houses which were either heated or lit by the new source of energy. By the mid-1920s, this figure had risen to one in seven, and, by the mid-1930s, to one in two. By 1937, five-sevenths of the total population of Manchester was probably benefiting from either an extensive or at least a 'basic' use of domestic electricity. Dissemination of the new energy source was closely related to class, income and housing. Those who lived near to the city centre, where large numbers of back-to-backs were still to be found, only rarely had access to domestic supply. Tenants or owners of the street upon street of terraced accommodation radiating outwards from the urban core towards the inner suburbs constituted a major market for the 'basic' electrical facilities of heat and light. Further out, in a rapidly expanding suburbia, large numbers of householders aspired towards a 'part'- or, among the genuinely affluent, an 'all'-electric house. Beyond the city boundaries, in 'progressive' overspill areas like the massive Wythenshawe estate, architects, builders and the council combined to encourage and even enforce the use of 'basic' electricity.[15]

It was middle-class suburbanites who led the way during the post-war electrical boom in 1920. There were said, early in that year, to be '2000 people waiting connection to the system and the Committee could not get the meters and the labour to connect them as quickly as it wished'.[16] But this was a period in which both gas and electricity were often still 'too expensive for general use'.[17] Addressing the same theme, the anonymous author of an article in the progressive and pro-electric *Manchester and Salford Woman Citizen* asked her readers in 1924 whether they were 'too conservative to abandon [their] kitchen range for a gas or electric cooker, or was it that we are afraid of gas and electricity "eating our money"?'.[18] With metered, flat-rate supplies still being charged at 6d per unit there could be little doubt as to which explanation carried the greater weight among potential working-class consumers.[19] The position was exacerbated following the phased introduction, during the earlier 1920s, of the 'all-in' system. This involved a fixed annual charge of 20 per cent of the 1914 rateable value of a house, plus a halfpenny per unit for lighting and other purposes.[20] Domestic supplies were certainly made more attractive to owners and long-term lessees of larger suburban houses, but to the great mass of working-class Mancunians, living in low-value back-to-backs, cheap 'weekly houses', or rented corporation accommodation, the scheme merely drew attention to the social exclusiveness of the new form of energy and the differential which separated the flat from the all-in rate.

From the perspective of the Electricity Department, the 'all-in' system promised to boost a dangerously stagnant domestic load – according to H. C. Lamb, the Department's long-serving chief engineer, only a tenth of the city's 200,000 houses had been wired by 1926.[21] What Lamb might have pleaded, by way of exoneration, was that from the early 1920s, the Department and its policies had been subjected to radical criticism from a powerful anti-municipal and 'anti-waste' faction in the Council. There was, for example, carping criticism in 1920 of the decision to enter into a new and more expensive contract with the then chief engineer, S. L. Pearce, who had been the driving-force behind the decision to go ahead with a massive and highly advanced power station at Barton.[22] Municipal trading and 'economy' debates provided attractive themes for Lord Mayors, when they made their autumn inaugural address before a full Council chamber and the local press. In 1923 W. T. Jackson, a moderate Conservative, cited Bernard Shaw and warned his audience that 'because a municipality provides music in the parks that was no reason why it should begin to manufacture trombones'.[23]

Such gnomic statements carried a clear enough message for those, on the Conservative benches, who drew a sharp distinction between the Department's right to demonstrate and its right to sell electrical appliances; and, when the embargo on the latter activity was finally broken in 1927, there were warnings that a door had been opened to 'extreme Socialism' and that independent traders were bound to be ruined during an economic downturn.[24] A typical Labour riposte during this controversy was that there were direct and illicit connections between individual councillors and powerful interests in the local electrical retail trade.²⁵ But, by 1927, anti-municipalist attacks on the Electricity Committee were beginning to give way to a more sustained and highly focused Labour critique. Answering progressive charges that existing pricing policies benefited affluent domestic and industrial users at the expense of poorer customers, William Walker, the long-serving chairman of the Electricity Department, insisted that 'if it were not for the big industrial users of electricity the small consumer would have to pay eightpence or ninepence per unit'.²⁶ Six months later, the Labour group, which was about forty-strong throughout the 1920s, and able to outvote the conservative majority if it could gain the support of Liberals and Independents, tried unsuccessfully to insist that wiring schemes should be installed by the department's own direct labour, rather than by outside contractors.²⁷

Then, at the end of 1928, the opposition pressed the claims of prepayment slot-meters, which would allow less well-off consumers to pay a relatively high unit rate to cover initial wiring costs, but a much lower one thereafter. Why, demanded the militant Tom Regan, who was to play a dominant role in the 'popular electricity' campaigns of the 1930s, was Manchester ignoring a scheme which had already proven itself in Sheffield, and which would do a great deal to even out the inexcusable differential between the price paid by working-class, corporation-housed consumers and larger domestic and industrial customers? On this occasion, also, the Labour group was defeated: but the accession of Liberal and Independent support had run the Conservatives close – the final motion was defeated by 54 to 44.²⁸ A year later, in December 1929, the 'cheap electricity' group, again skilfully marshalled by Regan, attacked the 'all-in' tariff and argued for an updating of the pre-war rating base together with concessions for smaller householders. Walker's response was classically legalistic and anti-socialist. 'If a reduction were to be made on behalf of any particular section of the community', he argued, 'that section would be receiving electricity which some one else paid for. It was illegal to make a reduction for any particular section of consumers at the expense of other consumers.' But significant numbers of Liberals and Independents had again been won over to the Labour cause and minor concessions were obtained for groups of less affluent consumers.²⁹

By 1931 both the Conservative majority and the Electricity Department were ready to acknowledge that electricity should be made more widely available. But there would be no capitulation. Prepayment meters were to be installed, but the 'minimum consumption per house per annum has been fixed at 100 kilowatt hours, and in order to take advantage of the scheme a consumer must agree to use not less than this amount. Installations will only be made when a reasonable number of applications is received in any particular locality.'[30] During the next two or three years, Labour attacks on the Department and the Conservative majority over the social implications of electricity diminished in intensity. Regan sniped at what he alleged to be excessive prepayment meter charges and the Department's over-harsh attitude towards corporation tenants who fell behind with their bills.[31] But this was a period of compromise rather than open conflict and Walker and Lamb were probably more exercised by attacks from outside Manchester: Hugh Quigley, the influential chief statistician at the CEB, accused Manchester of having cabled only 60 per cent of the streets within its administrative area.[32] Then, in the early summer of 1934, radicals and moderates were briefly united by a sudden attack on the underlying rationale of the modern municipal trading department. The Electricity Committee, no longer chaired by William Walker, and evidently embarrassed by the scale of its revived profits, unilaterally offered to donate £90,000 in aid of the general rate. Walker who had for so long encouraged the Department to operate as a profit-making and price-cutting, though essentially cautious and non-collectivist, municipal agency, emphasised the dangers of allowing local public utilities to be used as milch-cows to satisfy the demands of narrow-minded and regressive 'anti-waste' elements within the Council. He warned that the 'gift' would indefinitely delay lower prices for every category of consumer and alienate outside concerns which had relied for more than a decade on the Department to supply them with cheap current. Tom Regan went further. He tried to refer the issue back to the Electricity Committee and force it to use the £90,000 'gift' to make immediate and far-reaching price reductions – especially for hard-pressed groups of consumers. This ploy was defeated by 52 to 42 and a further amendment, to take the matter back to the Finance Committee, also went down by 48 to 41. Radical members refused to accept the finality of the vote and disrupted further proceedings. In their view the acceptance of the 'gift' marked a return to an individualistic, anti-municipalist and anti-socialist era. The sanctity and stability of the general rate was once again being allowed to override the financial needs of public services and of poorer members of the community. As for the Conservatives, their position was well summarised by the *Manchester City News*: 'Let us, at any rate, be glad that there is a surplus to benefit somebody, and rejoice in the outstanding efficiency of the Manchester electricity undertaking and the low charges of our electricity supply ... In the end it seemed that the Conservative benches, which carried the acceptance, weighed most the urgency of getting the rates down by any and every means.'³³

By late summer 1934, the unity engendered by the rejuvenated 'anti-waste' offensive had disintegrated and the radicals, led again by Tom Regan, were attempting to force the flat-rate service down from 4¼d to 4d a unit and claiming that Manchester's charges were 'becoming too heavy for the family purse'. But the Conservative majority was now more secure than it had been during the 1920s: a cheap power motion was defeated by 60 to 28; Regan and his supporters were accused of seeking 'to give electricity away to certain sections of the public'.³⁴ By 1935 the Department was priding itself on the fact that 'some 75,000 of the 200,000 houses in Manchester have electricity installed and 17,000 of the houses owe their electricity to the corporation's assisted wiring schemes or prepayment meters'.³⁵ Two years later it was asserted that 'extensive use of electrical equipment in the home is not a privilege which is within the reach only of the well-to-do, but, as regards cost and suitability, is one of which advantage can be taken by all householders'.³⁶ 'When the public generally becomes more accustomed to the most modern appliances, it is expected that ... timidity will disappear, with beneficial results to this [the less well-off] class of consumer.'³⁷

The 'electrical question' and its ideological ramifications, now came to prominence on the massive and already strife-prone 'progressive' estate of Wythenshawe. The leader of a community pressure group, J. E. Robinson, argued that since rents were significantly higher for new corporation housing than they had been in the older, inner-city districts from which many of the new tenants had come, a fixed charge for electricity of 6d a week was too high.³⁸ Residents complained that 'though in the dark months they have to pay extra because of their consumption of current being above the number of units for which the minimum charge is made [6d a week with the rent], in the lighter months when their consumption is below that scale, they are given no rebate by the department'.³⁹ When the issue was discussed in full Council, and a compromise reached whereby those living on corporation estates would pay a *minimum* of 26s a year rather than 6s 6d a quarter, William Walker, adopting a characteristically legalistic and 'business-like' stance, insisted that the radicals had succeeded in inflicting an enforced policy of subsidy on to the corporation. But Tom Regan was unrepentant. The Electricity

Department was neither as efficient nor as generous as its public statements implied; and 36 out of 40 municipal undertakings in London had a lower minimum charge than Manchester.[40]

Regan's radical critique lost little of its intensity with the approach of war. 'At the present time', he told the Council in July 1939, 'electricity is a luxury which many people can't afford. It costs about 2s a day to use an electric fire in a Corporation house and some of these fires have never been used at all except to see what they looked like when the tenant first moved in ... I believe a good deal of Manchester's sickness in winter is due to lowered resistance because people can't afford proper heating.'[41] In 1940 Regan told the Conservative majority that 'although we are putting electric radiators in our Corporation houses they are only being used in extreme circumstances, because electricity ... still remains a luxury for working class people. Every year many children have to go to hospital through the lack of heat in their bedrooms. What is the good of having life-saving appliances at our disposal if we refuse to apply them to those in need?'[42]

The Department had claimed in 1938 that 'under present conditions practically every householder living on one of the cable routes can have electricity installed with a minimum of disturbance to either his home or his savings'.[43] But for those – the two-sevenths, or nearly a quarter of a million – of Manchester's population which lived without access to electricity in the later 1930s, the outlook was less sanguine. The 'all-in' tariff, by far the cheapest and most economical category of supply, was still available only to the relatively affluent; the flat-rate charge, fixed at $3\frac{3}{4}$d for lighting, and $1\frac{1}{8}$d for 'multiple' use, had not declined as rapidly as the Department had hoped; and prepayment meters still cost $4\frac{1}{4}$d a unit. Most unsatisfactory of all, the assisted wiring scheme was based on an initial deposit of 19s 2d, followed by twenty quarterly payments of 9s 8d. Yet when a new consumer had found a niche among the bewildering range of schemes which ultimately reflected financial reliability and social status, the hire of basic electric essentials became more attractive. Cookers were expensive at 6s 6d a quarter, going down to 2s after three years but wash-boilers, at 2s 6d a quarter, and water-heaters at 2s 6d, were more reasonably priced.[44] All this strongly implies that reaching the first rung of the 'electrical ladder' in urban Britain in the inter-war period could be a complicated and dispiriting experience, but that, once 'accepted', the new consumer could gradually build up a basic repertoire of appliances and begin to derive tangible benefits from the new form of energy.

The force of the socialist critique of 'popular electricity' policies in Manchester in the inter-war period is nevertheless partially confirmed by the data in Table 3, which sets out the average price for 'private supply' and domestic units per head between 1935 and 1940 in six urban areas. (Data is available for an earlier period, but the distinction between industrial and domestic uses is not always clear.) What emerges is that the Manchester Department was probably leaving a very large proportion of its potential domestic market without an adequate supply. (The comparison with Leeds, which developed more slowly than Manchester in the 1920s but accelerated in the later 1930s is particularly revealing.)

**Table 3**  *Consumption of Electricity in Six Urban Areas: 1935–1940*

|  | Manchester | | | Leeds | | | Liverpool | | |
|---|---|---|---|---|---|---|---|---|---|
|  | *(i)* | *(ii)* | *(iii)* | *(i)* | *(ii)* | *(iii)* | *(i)* | *(ii)* | *(iii)* |
| 1935 | 855 | 1.20 | 99 | 485 | 1.19 | 146 | 1018 | 1.22 | 125 |
| 1936 | 855 | 1.15 | 112 | 487 | 1.12 | 161 | 1026 | 1.17 | 139 |
| 1937 | 855 | 1.11 | 130 | 487 | 1.05 | 189 | 1020 | 1.12 | 161 |
| 1938 | 855 | 1.06 | 144 | 490 | 1.03 | 202 | 1055 | 1.05 | 182 |
| 1939 | 855 | 1.03 | 158 | 492 | 1.03 | 214 | 1051 | 1.03 | 190 |
| 1940 | 855 | 1.03 | 165 | 494 | 1.03 | 215 | 1069 | 1.02 | 212 |

|  | Birmingham | | | Hackney | | | Southampton | | |
|---|---|---|---|---|---|---|---|---|---|
|  | *(i)* | *(ii)* | *(iii)* | *(i)* | *(ii)* | *(iii)* | *(i)* | *(ii)* | *(iii)* |
| 1935 | 1075 | 1.10 | 78 | 215 | 1.74 | 105(?) | 210 | 1.55 | 86 |
| 1936 | 1086 | 1.06 | 84 | 215 | 1.68 | 112 | 210 | 1.28 | 96 |
| 1937 | 1092 | 1.02 | 98 | 215 | 1.20 | 125 | 210 | 1.18 | 123 |
| 1938 | 1107 | 0.96 | 117 | 215 | 1.38 | 149 | 210 | 1.17 | 145 |
| 1939 | 1115 | 0.92 | 135 | 215 | 1.31 | 182 | 220 | 1.16 | 155 |
| 1940 | 1131 | 0.93 | 161 | 215 | 1.27 | 214 | 223 | 1.13 | 179 |

*Key*: (i) = Population of total supply area in thousands.
(ii) = Price per unit for domestic supply.
(iii) = Units per head of population: domestic supply only.
*Source: Electrical Times*, 'Annual Table of Electric Supply, Costs and Records (Public Authorities)'.

The Manchester Electricity Department was evidently strongly committed to increasing the size of its industrial and commercial loads and providing neighbouring companies with cheap power.[45] Whether these policies worked against optimal development of the domestic market can best be evaluated via an account of sales and advertising policies, and the technical and social attitudes by which they were shaped. The Department had a long and continuous record of advertising and 'electrical education' – as early as 1908 it had co-ordinated an exhibition consisting of no fewer than 320 stands.[46] But the post-war ethos tended to be technical and triumphalist – the Department would lead and an electrically ignorant citizenry follow on behind. The Department, readers of the city's official handbook were told in 1925, was 'probably the finest electricity enterprise in the country, but it is largely the creation of a few experts and interested laymen – the kind of benevolent autocracy that is often the best form of government in a democracy – and the general community cannot be said to have helped at all except by a blind acquiescence and a certain readiness to use the product that is offered to them'.[47] In terms of publicity, the Department set out to 'tempt the householder – as they tempted the manufacturer before him – with cheap current. Almost invariably the current first enters the house in the form of light, and occasionally ends up as a general servant – cooking, heating, hair-curling, bed-warming, washing, ironing and cleaning.'[48]

Here, as in other types of publicity, the Department committed itself to the triumphalist vision of houses and domestic routines transformed and made more 'modern' and elegant via the application of the new form of energy. 'The Committee', an anonymous author reflected in similar vein in 1927, 'is using every effort to tempt the householder to use electricity for the thousand and one services it normally fulfils in America.'[49] But Manchester was emphatically not New York and there was more than a little wish-fulfilment in the Department's claim in 1928 that the public 'is recognising, more and more, the infinite capacity of electricity for service in the home'.[50] When a journalist from *The Electrician* journeyed north in the autumn of 1930, he reported that the corporation was convinced of the need to prepare the public for the return of better and more affluent times. A film depicting 'aspects of electric supply from the power station to the home' had been produced, and a great deal was expected of the all-in tariff, the assisted wiring scheme and hire-purchase of cookers and wash-boilers. A spokesman was convinced that the recently introduced convention of adding a fixed amount for electricity on to weekly rent was beginning to make electricity more popular with those who lived in the 'older type of house'.[51] By 1931, the Department was producing larger numbers of posters directed at those not wealthy enough to subscribe to the all-in tariff.[52]

In newspaper advertisements potential consumers were told that 'there is no cleaner or better means of cooking than by electricity. With clock-like precision, an electric cooker performs all the duties of the kitchen range, but with an entire absence of soot, smell and flame.'³³ Yet the rate of adoption among the mass of the population remained disappointingly low. 'Saturation point in the matter of electricity supplies', an anonymous member of the Electricity Committee commented in 1932, '... is very far from being reached, and ample scope remains for further development'.³⁴ Elaborating on this theme, a Department spokesman told a journalist in 1933 that 'there is still a certain diffidence on the public's part in making real use of the electricity supply when they have got it'. But when this line of argument was developed further, it became clear that greater attention was still being given to the affluent, 'modern', all-in consumer than to those who rented 'weekly houses'. 'Once they have obtained an electricity service under the all-in domestic tariff', the spokesman enthused, 'we want them to realise all its possibilities not only in the well-known features of the electric cooker, the radiator, the vacuum cleaner, the sewing-machine motor, the electric iron, and so on, but in the newer features also of the infallible electric clock, the all-mains radio and gramophone, the electric water heater ... These creations of the new age must not be neglected in our city.'³⁵ This was the climate in which the Department preached to the converted – the relatively affluent owners of electric cookers – and organised lectures and displays on 'milk cookery', 'cold cookery', 'Christmas cookery' and, when the time was ripe, 'economical war-time cookery'.³⁶

By this date, also, electrical 'shows' were regularly touring the wealthier suburbs. At Chorlton Baths in the southern suburbs in 1935, a centrally placed floral arrangement, surrounded by wooden chairs, was flanked by cubicles, each of them prominently marked 'Home Aids', 'Heating', 'Lighting'; the centre-piece of the entire display, an 'Electric Home' was emblazoned with the Department's logo, 'MCED – Power, Heat, Light'.³⁷ If suburbanites were tempted by this array of electrical appliances and services, and committed themselves to a purchase or to hire-purchase, a lady demonstrator would immediately call to explain how to achieve the best results.³⁸ Whether at 'displays', showrooms, shops, or in private homes, this was the golden age of 'demonstration'.

The dominant ethos of the Electricity Department in the later 1930s was captured by another journalist who reported that 'High-pressure salesmanship is not favoured ... [the Department] prefers to rely upon well thought-out schemes of tariffs and hiring services, and reasoned argument. It is considered that good service cannot be offered where over-selling is permitted, and also that it is unjust to permit one class of consumer to be subsidised by another class."³⁹ There was a heavy emphasis on industry – engineering, rubber manufacture, spot welding and thermal storage heating were important areas and novel applications were regularly demonstrated in the Department's workshop.[60] But perhaps the proudest boast of all was that 'all corporation dwellings, even when gas services are provided for cooking and washing, are now wired with a small plug for a portable appliance and [there are] heating plugs in bedrooms.[61] If the Department's aims and approach were now marked by a degree of 'democratisation', nowhere was this more vividly symbolised than in the new showrooms which were opened in the Town Hall in 1938 – a central venue now shared with the great competitor, gas. Here in the main hall, a 'multi-coloured illumination is thrown on to the spray [of a fountain] . . . by means of two rotating drums containing electric lamps and sections of coloured glass'.[62] Elegant standard lamps were arranged around a central, circular bed of mock-natural flowers: and to the far left and right were ranged electrical appliances for sale. White, symmetrical paving and shiny metallic banisters directed the eye towards a stairway and a lower floor, consisting of a lecture theatre, a workshop and a 'room devoted to illustrations of the possibilities offered by modern electric lighting'.[63] The most youthful of the trading departments had finally found its rightful home; Cottonopolis had embraced art deco.

In terms of imagery and self-presentation, then, 'beautification' and 'modernity' outweighed a more earthy and popularising municipalism. In a pamphlet *Hot Water by Electricity*, published in 1938, a prototypically modern and 'elegant' woman was depicted reading by a standard lamp, washing her 'delicate' underwear, and getting ready to step into a marbled bath; the significance of every action was underlined by Samuel Butler's ironic words in *The Way of All Flesh*, 'We have all sinned and come short of making ourselves as comfortable as we easily might have done'.[64] In *Refrigeration by Electricity* the reader was asked to recall a time 'when you "didn't need" a 'phone in the house. Today you wonder how you ever managed without it. Or if it wasn't a 'phone, it was Electric Light, or a Car; or your first Wireless Set, or the Vacuum Cleaner.' Now, it was implied, a refrigerator and 'cold cookery' were equally indispensable.[65] Only in the less ambitious format of press advertising were the basic 'electrical essentials' – how much 'cleaner, healthier, happier, safer', the new source of energy had proven itself to be – given strong emphasis.[66]

Confronted on one side by the anti-municipal and 'anti-waste' faction, and, on the other, by a socialist vision of the new source of power as a potential free good and subsidised public service, the Manchester Electricity Department opted during these years for what it believed to be an apolitical *via media*, based on moderate municipalism, the bed-rock of an expanding industrial and commercial load, and a broadly middle-class publicity programme for domestic supply. But each of these responses was conditioned, to a greater or lesser extent, by the continuing presence, and, in some instances, the economic and cultural supremacy of gas. Founded as early as 1824 the Gas Department had quickly established itself as the premier trading department in the city. By the later nineteenth century, however, this very primacy had drawn the Committee into a complex and largely sterile debate: the balance to be struck between a cheap supply and the proportion of profits to be handed over in support of the general rate.[67] In the short term, an exceptionally buoyant demand more than compensated for artificially inflated prices, low levels of investment, and minimal technological innovation. The introduction of slot-meters in 1890 was particularly successful, and, by the early twentieth century, Manchester had indisputably entered a golden age of gas.[68] It was the outbreak of war, rather than competition from electricity, which suddenly threatened to bring the industry to its knees. High levels of domestic connections continued to be sustained – by 1917 most homes in the city were in a position to receive a supply – but shortages of coal, inadequate transport facilities, and the continuing requirement to subsidise the rates were undermining the Gas Department's long-term viability.[69] In 1920 the trading departments were relieved of their duty to support the general rate, and this finally cleared the way for the establishment of a reserve and investment fund. But the ever-spiralling cost of coal, together with shortfalls in revenue with which to fund post-war wage demands, led to the widespread and, as we shall see, mistaken view that the Gas Department was failing to respond to the 'challenge of electricity'.

Full crisis was precipitated by the Bradford Road Station explosion in September 1927. The Gas Department found itself indicted by municipal and governmental investigators, but no sooner had technical and organisational reforms begun to bear fruit in the early 1930s than a second series of explosions, in October 1933, further eroded the credibility of Manchester gas. The *Daily Express* coined the mocking phrase 'City of Civic Muddle' and the accusation that the older utility was capitulating to electricity was once again widely canvassed.[70] But there were discrepancies between public perceptions of gas in Manchester during these years, and its actual performance. Statistical evidence is scarce and ambiguous, but it seems certain that the Electricity Department was insufficiently experienced and committed to the rapid expansion of cheap domestic supplies to capitalise on its competitor's misfortunes and mistakes. From the mid-1920s, also, the Gas Department clearly drew selectively and creatively on sales and design strategies which had already been developed on a narrow front by its rival. The gas authorities were able to boast, in 1925, that 'the hire of silent, artistic hygienic gas fires for a few shillings a year is doing much to expand the ever-growing business of the department'.[71]

By 1928, in the aftermath of the Bradford Street explosion, a 'definite sales policy' had already been formulated.[72] Three years later a wide range of appliances could be seen on display at a showroom in the burgeoning southern suburb of Withington, and a newly designed cooker, in which a single flame at the back had replaced a burner on each side, had come on to the market.[73] In 1932 a personal follow-up service to every consumer was said to be further boosting domestic sales.[74] And by 1933 it was being claimed that 'gas fires can be finished with any colour however delicate the shade desired may be . . . It is thus possible to provide a fire to harmonise with any furnishing colour or period scheme. The gas fire therefore can in addition to its utility become a definite and attractive part of a harmonised colour scheme.'[75] More than a little had clearly been learnt from triumphalist rhetoric depicting an energy source as a potential transformer of life-style and domestic ambience. Gas was less 'old-fashioned' and less rooted in a redundant and penny-pinching genre of municipalism than its 'modernist' critics allowed.

By autumn 1935, the Gas Department had introduced its own version of the all-in tariff and devised a campaign to encourage rehoused slum-dwellers to opt for the older rather than the new form of energy.[76] 'Hardship is being inflicted on people who are moved to these flats and who are forced to use electricity for lighting, cooking and heating', a member of the Gas Committee complained in 1936. 'They are unfamiliar with it, they cannot afford it, and they resent being dictated to. Their ideas of economy are outraged, for they are accustomed to penny-in-the-slot payments and turning lights full on or only half on as circumstances suggest.'[77] 'By what authority', demanded a tenant in the same year, 'can the City Council dictate that we must not use gas?'[78] Gas had history and the slot-meter on its side. More important, at 10d a therm, it was still significantly cheaper for the smaller consumer.[79] By the end of the decade, the Gas Department had regained its poise and commercial acumen following the traumas of the 1920s and early 1930s. Located next to each other in the Town Hall extension, 'the electricity and gas industries rivalled one another in the way in which they extol their services by means of well equipped showrooms, window displays, demonstration rooms and lecture theatres'.[80] 'Perhaps the most striking thing', it was stated in 1938, 'is that gas is still so far ahead of electricity'.[81] This was an exaggeration, but it serves as a warning that neither in Manchester, nor, probably, in any other major British city, did electricity in any sense 'replace' gas in the domestic sphere in the inter-war era, or in the period after 1945.

## Notes

1  *ET*, 20 April 1922, 379.
2  Margaret Partridge, 'Women Co-operators and Electricity', *ET*, 29 January 1925, 141–2.
3  *E*, 16 July 1926, 62.
4  Ibid, 3 June 1927, 600.
5  *EII*, 30 January 1929, 162.
6  *ET*, 20 August 1931, 271.
7  Ibid.
8  *E*, 22 June 1934, 842. By 1938, 60 per cent of all houses in Glasgow were said to be wired ('EAW at Glasgow', *E*, 26 June 1938, 803). Comparative data for Leeds can be gleaned from W. G. Rimmer, 'Leeds and its Industrial Growth: – Gas and Electricity (ii)', *Leeds Journal*, 28, 1957, 299–303.
9  There is support for the view that, in the North, the domestic load was developed less rapidly than industrial and commercial supply in Sue M. Bowden, 'The Consumer Durables Revolution in England 1932–1938: a Regional Analysis', *Explorations in Economic History*, 25, 1988, 57.
10  E. E. Edwards, *Report on Electricity*, 31.
11  Ministry of Transport: *Report of the Committee on Electricity Distribution* (HMSO, 1936), 64.
12  *ER*, 19 October 1923, 562; and *E*, 30 June 1922, 771.
13  Some of these issues are illuminated on a broad, comparative scale by Thomas P. Hughes in *Networks of Power: Electrification in Western Society 1880–1930* (Baltimore, 1983). Case studies have been provided by Mark H. Rose and John Clark, 'Light, Heat and Power: Energy Choices in Kansas City, Wichita and Denver, 1900–1935', *Journal of Urban History*, 5, 1979, 340–64; and Edmund N. Todd, 'A Tale of Three Cities: Electrification and the Structure of choice in the Ruhr, 1886–1900', *Social Studies of Science*, 17, 1987, 387–412. There is no comparable work on Britain but I. C. R. Byatt has laid the foundations for the early 'pioneering' period in *The British Electrical Industry 1875–1914: the Economic Returns of a New Industry* (Oxford, 1979).
14  This account is based on Arthur Redford, assisted by I. S. Russell, *The History of Local Government in Manchester. Vol. III: The Last Half Century* (1940), 94–111; and *How Manchester Is Managed* (1936). H. H. Ballin, *The Organisation of Electricity Supply in Great Britain* (1946), chap. 2, provides succinct background on municipal trading and electricity.

**15** This outline should be compared with the overview provided in John Burnett, *A Social History of Housing 1815–1970* (Newton Abbott, 1978), chap. 7.
**16** *ER*, 2 January 1920, 15.
**17** E. D. Simon and Marion Fitzgerald, *The Smokeless City* (1922), 56; see, also, their comments on p. 61 of the same book. As late as 1929, *The Times* stated that 'it is only the well-to-do people, who have to consider the wages of domestic service, who occasionally use electricity for the purpose of heating a room' (31 October 1929).
**18** 'Let There Be Light', *Manchester and Salford Woman Citizen*, 15 November 1924. See, also, ibid, 15 February 1923.
**19** City of Manchester: Appendix to Council *Minutes*, 2 July 1924, 469.
**20** *How Manchester Is Managed* (1926), 46.
**21** *ER*, 10 October 1926, 526. For biographical detail on H. C. Lamb see *Manchester City News*, 2 July 1938.
**22** *Manchester Guardian*, 5 August 1920; and *E*, 13 August 1920, 172.
**23** *Manchester City News*, 10 November 1923.
**24** *ER*, 14 January 1927, 60; and *Manchester City News*, 8 January 1927. The Manchester Retail Trades Association canvassed the Council to prevent the Department from selling, rather than merely renting, apparatus. See *ER*, 18 February 1927, 257.
**25** *ER*, 13 August 1926, 259, citing the *Daily Dispatch*.
**26** *Manchester City News*, 9 July 1927. Sir William Walker (1868–1961), a director of Henry Simon Ltd, served on the Electricity Committee for 39 years, 6 of them as chairman. He was appointed to the CEB in 1926 and was later a part-time member of the Central Electricity Authority between 1948 and 1952. The national salary scale for chief electrical engineers, which Walker devised, was named after him. (See W. E. Swale, *Forerunners of the North Western Electricity Board* (Manchester, 1963, 76.)
**27** *Manchester City News*, 7 January 1928.
**28** Ibid, 8 December 1928. Tom Regan is an important though under-documented figure in the history of Mancunian radicalism. The originator of the city's now-booming Ringway Airport, he became a controversial Lord Mayor in the 1950s. He was an anti-monarchist and visitor to the Soviet Union during the coldest years of the Cold War. In his retirement he compiled the useful *Labour Members of the City Council 1894–1965* (1966), a manuscript copy of which is available in Manchester Central Reference Library. For biographical details, see *Manchester Evening News*, 13 September 1976.
**29** *Manchester and Salford Woman Citizen*, 20 December 1929, 11. On the issue of the rating base, see J. W. Maitland, 'Our Electricity Supply: Charges for Electric Current', *Manchester City News*, 21 September 1929.
**30** *How Manchester Is Managed* (1931), 52.
**31** *Manchester and Salford Woman Citizen*, 20 January 1931, 11 and 21 November 1932, 15.
**32** *Manchester Evening Chronicle*, 2 November 1933. There were also complaints at this time about muddle and inefficiency over the change from DC to AC. See *Manchester Evening News*, 28 December 1933.
**33** *Manchester City News*, 9 June 1934. The full story of the 'gift' can be gleaned from *Manchester Evening Chronicle*, 17 May 1934, and *Manchester and Salford*

*Woman Citizen,* 20 June 1934, 14.
34  *Manchester Guardian,* 5 July 1934.
35  Ibid, 7 August 1935.
36  *How Manchester Is Managed* (1937), 87.
37  Ibid, 88.
38  *Manchester Evening News,* 31 March 1937. For further details of the Wythenshawe campaign see ibid 1 April and 19 April 1937, *Manchester Evening Chronicle,* 19 April 1937 and *Manchester City News,* 4 June 1937. The domestic economy of Wythenshawe during this period has been splendidly described by the Manchester Women's History Group, 'Ideology and Bricks and Mortar: Women's Housing in Manchester between the Wars', *North-West Labour History,* 12, 1987, 12, 24–48. The municipal context is briefly outlined by Jennifer Dale in 'Class Struggle, Social Policy and State Structure: Central–Local Relations and Housing Policy' in Joseph Melling (ed.), *Housing, Social Policy and the State* (1980), 194–223.
39  *Manchester Guardian,* 3 April 1937.
40  *Manchester and Salford Woman Citizen,* May 1937, 13.
41  *Manchester City News,* 8 July 1939.
42  Ibid, 10 February 1940.
43  *Homework Made Easy* (1938), 4.
44  Ibid, 4–10.
45  Redford and Russell, 272–3.
46  Swale, 69.
47  *How Manchester Is Managed* (1925), 16.
48  Ibid, 16.
49  'The Electricity Age: Manchester's Lead to Britain', *Manchester City News,* 17 September 1927.
50  *How Manchester Is Managed* (1928), 62.
51  'Autumn Activities in Manchester', *E,* 26 September 1930, 369.
52  *ER,* 2 October 1931, 507.
53  'Cook by Electricity'. Advertisement in *How Manchester Is Managed* (1932), 202.
54  Ibid, 209.
55  *Manchester City News,* 21 October 1933.
56  *Manchester Guardian,* 14 May 1935; 7 July 1936; 14 December 1938; and 20 January 1940.
57  This account is based on a photograph in *ER,* 22 November 1935, 719.
58  *How Manchester Is Managed* (1936), 208.
59  'Development Activities in Manchester', *ER,* 22 January 1937, 123.
60  Ibid, 124. The Department's industrial campaign was supplemented by its publication, *Power Service Bulletin.*
61  *ER,* 22 January 1937, 124–5.
62  'Extensions at Manchester', *E,* 13 May 1938, 601.
63  *How Manchester is Managed* (1939), 84.
64  *Hot Water by Electricity* (1938), 1.
65  *Refrigeration By Electricity* (1938), 1. In 1939 the Department also emphasised that 'refrigeration is advocated as an all-the-year-round advantage ... [but] it is still regarded by the public as a hot weather luxury'.

66 Advertisement printed in *Manchester and Salford Woman Citizen*, May 1939.
67 The following account of the Gas Department is based on Redford and Russell, 77–93 and 274–81.
68 Shena Simon, *A Century of City Government*, 363–5.
69 Redford and Russell, 90.
70 Ibid, 279.
71 *How Manchester Is Managed* (1925), 11.
72 Ibid (1933), 204.
73 Ibid (1932), 193–4.
74 Ibid (1933), 204.
75 Ibid (1934), 221.
76 Redford and Russell, 280–1.
77 *Manchester City News*, 17 January 1936.
78 Ibid.
79 This figure has been derived from the reports of the Gas Committee included throughout the 1930s in *How Manchester Is Managed*.
80 'Manchester City Hall', *E*, 27 May 1938, 696.
81 Shena Simon, 371.

# 5
# Rural stagnation

In no sphere of electrical development did the rhetoric of triumphalism fall on stonier ground than in the British countryside. The rate of village and farm connection may have been higher in the 1930s than in the 1920s, but, compared with France, Germany, Scandinavia, Japan and many states in the USA, the record was not an impressive one. Agrarian and planning activists were convinced that the new source of energy had a crucial role to play in stimulating arable production via scientific electro-culture, restoring the vitality of the traditional 'organic' village community and galvanising rural crafts and industries. During the 1920s electro-culture, under the enthusiastic leadership of the electrical engineer and journalist Richard Borlase Matthews, exerted considerable influence.[1] But when, during the later 1920s and early 1930s, increasing numbers of agricultural experts cast doubt on the efficacy of applying electric current directly to arable and other crops, disenchantment set in.

It was at this juncture that electro-culturalists began to throw their weight behind the longer-established rural revivalist movement and insisted that the mechanisation of existing productive processes, and of traditional village crafts and skills, held the key to full economic recovery. The explicitly cultural goals of the movement proved to be chimerical, but electricity was nevertheless applied to an ever-wider range of farming activities, old and new. These included the incubation of day-old chicks, the sorting of eggs, canning and storage of fruit and vegetables, milking, sheep shearing and chaff cutting. Such developments owed little to the utopianism of the electro-culturalists and confirmed the existence of a wide gulf between the incessant flow of triumphalist propaganda, which insisted that British agriculture was lagging behind best practice in Europe, and the realities of technological innovation in the different farming regions. Nor could this tension be resolved by an appeal to quantitative evidence. International comparisons were, and still are, unreliable; and the indexes deployed by triumphalists – the percentage of the total non-urban population having access to 'basic electrical services', the numbers of villages receiving full supply, the proportion of isolated farms making use of the new source of power either for productive or domestic purposes – invariably referred to subtly different aspects of social life. Yet even when the statistics are partially reworked and due allowance made for the exaggerations and *non sequiturs* canvassed by triumphalists and rural revivalists, electrification in non-urban areas in inter-war Britain still seems to have proceeded at a slow and uncertain rate.

'We in this country', complained the *Electrical Review* in 1922, 'are far behind the Continent, where the application of electricity has attained an extraordinary degree of development'.[2] Forty thousand farms, it was estimated two years later, were large enough to 'profit materially by the aid of electricity', but little was being done by way of official encouragement.[3] In 1925 Lloyd George stated that Britain was the 'only country where there is no systematic effort being used to bring electricity to the aid of agriculture',[4] and a year later Borlase Matthews reported that only 500 farmers were using the new source of energy for productive purposes.[5] That this was an exceedingly small proportion of the total could be confirmed by reference to the performance of other countries. In Germany, in 1926, 90 per cent of all farms were estimated to have access to electricity. In France there would soon be an 'extensive agricultural electrification'. In America farm connections for mainly domestic purposes were running at roughly ten times the British rate, and in Japan electricity was being applied to the drainage and irrigation of rice fields, the husking of rice, silkworm rearing and moth killing.[6] Measured against achievements such as these, British enthusiasm for the stimulation of greenhouse plants, and the replacement of the horse by an as-yet-uninvented electric plough seemed misdirected. So did hopes for a larger rural domestic load when, according to *Electrical Industries and Investments* in 1929, the 'market-towns of England have hardly been touched'.[7]

By 1929 Herbert Morrison, as Minister of Transport, was telling a sceptical House that 'over 70 per cent of the rural population is at present included within areas with an authorised supply of electricity'.[8] But members representing agricultural regions insisted that the new and imposing lines of pylons were not yet delivering light and heat to village communities. Many local supply companies were unwilling to risk involvement in financially risky rural schemes. 'Like the ship-wrecked mariner', claimed the rural electrical activist, Sir Douglas Newton in 1930, 'it is a case of "Water, water, everywhere and not a drop to drink"'.⁹Borlase Matthews complained in 1932 that no more than about 4000 out of a total of nearly 420,000 farms were able to make use of electricity for productive purposes.¹⁰

A year later, in a wide-ranging and authoritative address to the Incorporated Institution of Municipal Engineers' Association, H. J. Denham placed blame on the self-reinforcing timidity and inactivity of the government and the rural supply companies. 'It is surely not illogical', he said, 'to suggest that a company which leads the world in the quality of its civilisation and of its social services, could make the one positive step in its electrification which would make the national grid finally worthy of its name, without destroying the enterprise of its constituent suppliers and distributors, and the prospect of a remarkable return on their invested capital.'¹¹ New marketing requirements, particularly for milk, now began to elicit a higher degree of product standardisation, thus persuading a small though significant minority of farmers to make long-term investment in the electrification of machinery.¹² But, as late as 1935, 'only about 6500 farmers out of a total of 395,800 were using electricity for productive purposes'.¹³ A year later, Sir Douglas Newton (now Lord Eltisley) claimed to be able to detect a change of mood. Farmers were disillusioned with or ignorant of the blandishments of electro-culture, and unimpressed by urban publicity which consisted of 'picturesque posters displaying lovely ladies in luxurious surroundings'. Once they had obtained electricity for their own homes, they were more willing to experiment with productive applications. And it was this process, so Eltisley argued, which underlay the fact that over 25,000 farms had now been connected up to the mains.¹⁴ (The large discrepancy between Matthews's and Eltisley's figures is accounted for by the former's omission of the exclusively 'domestic' category.)

Complaints, both inside and outside Parliament, that the government was continuing to ignore the electrical needs of the farming community and that rural consumers were being forced to pay monopoly prices, continued to be heard.¹⁵ Sir Frank Sanderson told the House in 1938 that farmers in East Sussex were being asked for a down payment of £800 before they were connected; and Sir Rupert De la Bere, the member for Evesham, inveighed against the 'exploitation of the farmers by the electricity companies'.¹⁶ 'District Engineer' painted a depressing picture in 1939 of an anonymous and supine supply company, with responsibilities in town and country, which put all its energies into developing a proven urban market and refused to venture out into the countryside. Such undertakings, 'District Engineer' went on, were always ready to hide behind the Electricity Commissioners' stringent safety regulations and rely on local contractors, rather than their own salesmen, to identify a financially viable rural demand.[17] Pressure and criticism were sustained until the outbreak of war but the government insisted that supply was now theoretically available to very nearly every member of the rural community and that it was the absence of a properly co-ordinated demand, rather than laxness on the part of supply undertakings, which underlay the failure to translate promise into reality. This was the climate in which De la Bere's Rural Electrification Bill, based on the assertion that '90 per cent of agricultural holdings remain without electricity supply of any kind', was defeated, without a vote, in April 1939.[18]

Conditions were worse in Scotland, Wales and Ulster. By the late 1920s the Clyde Valley Electric Power Company had established a number of joint rural–urban schemes[19] and, during the 1930s, the Fife Power Company extended 12,000 volt mains into country areas, before transforming down to 440 for local distribution.[20] But it was the thousand-square-mile Dumfriesshire scheme which best demonstrated the potential for rural electrification for Scotland as a whole. With a total potential market of about 15,000 consumers, the county-council sponsored plan provided for a domestic supply for about 750 in 1933, just over 4500 in 1935, and about 8000 by 1938.[21] But this initiative was exceptional and throughout the later 1930s, there were complaints in Parliament that Scottish companies were asking exorbitantly high premiums to connect remotely located premises, and ignoring the electrical needs of farmers and crofters.[22]

As for Wales, it was authoritatively stated in 1929 that 'more than half Carmarthenshire, the whole of Pembrokeshire, more than half Cardiganshire and the whole of Brecon and Radnorshire have not been provided for at all'.[23] In the same year, the Minister of Transport, Wilfred Ashley, echoed a common attitude when he stated that 'it always seems to me to be so uneconomical to run a line up . . . a Welsh valley, however nice the sun may shine over the Welsh hills, for you cannot tax the people . . . in order to provide electricity there'.[24] There was limited progress in South Wales, with just over 4000 consumers out of a total of 5000 receiving a domestic supply under a local scheme by the later 1930s.[25] But, as late as 1938, rushlights and tallow-dip candles were often encountered in Welsh villages and hamlets, while in more accessible areas, wax candles and paraffin were still the main forms of lighting. Large-scale electrification might, as one commentator argued, have helped to break down the economically disruptive interactions between high levels of out-migration from rural areas and urban and industrial decline. But there was no national or regional blueprint to encourage public supply to radiate outwards from market towns; and numerous farmers who had been connected up tended to minimise their financial gains by using current too sparingly and for too limited a range of domestic and productive purposes.[26] If, in terms of kilowatt consumption per head per annum, rural Wales was backward, Ulster had scarcely emerged from the dark ages. The figure there in the mid-1920s was a miserly 43 kilowatts, compared with Britain at 140, and the hydroelectricity-rich Switzerland and Norway at 900 and 2500 respectively. Data is not available for Ulster for the 1930s but, in 1937, there were still only very small numbers of consumers outside Belfast. (Eire was even more deprived: per capita consumption in the 1920s was 16 kilowatts.)[27]

When contemporaries sought explanations for the relative backwardness of electricity supply in the British countryside, they invariably drew attention to the diversionary impact of 'scientific electro-culture'. This, as we have seen, involved the application of small amounts of current to arable crops, fruit, flowers and vegetables, and had first come to prominence in the early 1920s. Even at this time sceptics were concerned both about the diversity of scientific results and the length of time separating investigation from large-scale application.[28] Others were convinced that progress would have been more rapid, had research been co-ordinated by the electricity industry rather than being pursued by academics working at Imperial College.[29] Stressing the interactions between electro-cultural and labour-reducing functions, Borlase Matthews pushed ahead with investigations in East Sussex and publicised the activities of other practitioners like the pioneer of the British electric plough, Major McDowell.[30] Matthews also canvassed the use of ultra-violet to hasten the rate of growth of flowers and vegetables. 'Daffodils and Lent lilies', he reported from the second British Electro-Farming Conference at Reading in 1926,

> when placed under the light of 1000 W Mazda lamps and reflectors for six hours a night, flowered in three days, growing about three-quarters of an inch a day. Narcissi flowered in seven days. The control plants, placed away from the light took four weeks to flower. Further experiments had shown that when transplanted seedlings were exposed to extensive illumination for one night they did not wilt, but actually put on a week's growth and were exceptionally strong and healthy.[31]

It was horticultural 'miracles' such as these which convinced Matthews and others that a revolutionary form of electro-culture would in the near future be underwritten by academic research and thus hasten the full electrification of British agriculture. 'The passage of a small electric current through the soil', he wrote in 1927, 'may to-day seem a somewhat futile and useless experiment, but in the days of our children's children such procedures may be an everyday commonplace part of cultivation'.[32] But by 1930 a specialist on the scientific status and future potential of electro-culture argued that the term now covered several distinct and potentially confusing aspects – the application to plants of a static discharge at high potential and frequency; a low-pressure galvanic current transmitted from a metallic network embedded in the soil; and the production of artificial light to hasten the growth of branches and foliage above ground.[33] In a survey published by the EDA in 1932 the claims of the more utopian of the electro-culturalists received firm rebuttal: 'It has been found that cereal crops do benefit . . . but the resulting increased yield was so small that practical application on the commercial scale is at present out of the question.' The hypothesis, in other words, that crops could be electrically 'treated' in order to boost production and profits still remained little more than a hypothesis.[34] Borlase Matthews rapidly adjusted to the spirit of the times: his journal *Electro-Farming* became *Rural Electrification and Electro-Farming* and gave increased coverage to the domestic uses of electricity in extra-urban areas. In 1934, Sir William Ray, who had recently been appointed director of the EDA, published what could only be interpreted as an obituary for electro-culture in its classic form: 'Instead of being offered as a handy means of driving farm machinery or of lighting farm buildings', he wrote, 'electricity was first put forward as a stimulation of crops through high tension discharges in the air or currents in the soil. Electro-culture has not yet got beyond the experimental stage and the disappointment that followed the rather sanguine claims once made for it was a handicap when the really useful workday applications of electricity came to be put before farmers.'[35] Electro-culture, he went on, meshed in with a set of beliefs which exaggerated the extent to which the new source could be expected to transform day-to-day work on the farm. 'Another handicap', Ray concluded, 'was the rather revolutionary picture drawn by some people in the first flush of enthusiasm. According to this picture, farmers ought to scrap all their existing methods and go in for a kind of Robot equipment.'[36] This was a reference to the electro-culturalists' advocacy of a new and quasi-automated 'scientific' agriculture and their insistence that every farmer must transform himself into a businessman capable of estimating the exact cost of a unit of production and ready to replace horses and men with electrically driven motors and tractors. Ray underlined the tendency, inherent in every branch of triumphalist theorising, to abstract electricity from its social context, and to depict it as a suprahuman *force* capable of revolutionising every existing productive routine.

Wayleave policies were also attacked by those who were convinced that rural electrification in Britain was dangerously retarded. 'Are we to lag behind?', demanded the *Electrical Review* in yet another editorial which made disparaging comparisons between Britain and Europe. 'Shall we let the wayleave problem block the way?'[37] 'This is the only country in the world where undue obstacles, regulations, and expenses are allowed to retard or prohibit the running of overhead lines.'[38] In 1926, the *Electrical Review* again reminded its readers that 'the supply of electricity is a public service, which must not be subordinated to the whims of private persons or the way-leave cupidity of the owners and occupiers of land. We do not suggest that electrical undertakers should be granted autocratic powers or ride roughshod over private rights; but they should not be at the mercy of every profiteer that they encounter, nor should they be required to wait for months before wayleaves can be obtained.'[39]

The Electrical Development Association might plead in 1927 for consensus rather than conflict over so potentially incendiary an issue,[40] but triumphalists, realising, perhaps for the first time, the potential strength of localised opposition to the National Grid, adopted a more forceful stance – 'we are going to have cheaper overhead lines whether the opposition likes it or not'.[41] Borlase Matthews was convinced that progressive farmers ought to be ready to pay, rather than be paid for wayleave rents.[42] 'We live in a world with others', he wrote in 1929, 'and neighbourly forbearance can do much toward making this world a better place to live in. The selfish . . . attitude is one to be deplored, and one which retards progress, and in the end does harm to the community as a whole.'[43] Margaret Partridge was also committed to negotiation rather than confrontation and, in a half-factual, half-fictional account of wayleave wrangling, written in 1930, she tactfully depicted a seemingly conservative squire as being less 'grasping' than an urban-based 'public body' which had recently bought up a country seat.[44]

Wayleave 'resistance' actually played no more than a minor role in delaying the completion of the Grid or the erection of local lines unrelated to the national network.[45] But pro-electric rural activists remained convinced that the absence of a formally negotiated national scale complicated and retarded electrical development throughout the 1930s.[46] 'In this country more, perhaps, than in any other', concluded a report on this topic in 1938, 'secondary difficulties such as the high standard of construction laid down by the Electricity Commission, opposition to the erection of overhead lines on aesthetic grounds and obstruction in the matter of gaining wayleaves, have helped to prevent a more general penetration of electricity supply in the countryside.'[47]

Yet too intense a preoccupation with the wayleave problem or with the alleged conservatism of the supply industry can divert attention from the positive achievements of a minority of 'progressive' rural electrification schemes.[48] Early experiments – notably in Herefordshire in 1922 – had been undermined by weak administration and national economic recession.[49] By the mid-1920s, valuable lessons had clearly been learnt and several predominantly urban companies had begun to venture outwards into country areas. In 1924 S. E. Britton spearheaded the electrification of a rural 'show-place' in Cheshire, and by 1933, 49 per cent of an area containing just over 6000 premises had been connected up.[50] The economic and domestic advantages to the farming community – in terms of heating and lighting, the mechanisation of 'traditional' tasks and the introduction of new techniques in dairy and poultry farming – were clear enough.[51] The success of the Chester project and a small number of similar schemes created a climate in which parliamentarians and academics gave increased attention to the problem of rural electrification – what could be learnt from other, and more successful European countries, and how, and on what scale, the state should be encouraged to sponsor its own initiatives.

Following a visit to Sweden in 1927, the Conservative Parliamentary Agricultural Committee recommended that the government should set aside funds for the electrification of rural areas by approved bodies; that the Electricity Commissioners should make themselves better acquainted with agricultural affairs; and, that 'rural experimental areas' should be set up.[52] Under pressure from the Ministry of Transport, the Commissioners announced the establishment of a demonstration scheme which would link Bedford to its surrounding hinterland. The aim of the project, which comprised an area containing just over 4000 premises and approximately 16,000 people, was to achieve 75 per cent electrification of all dwellings within two years by means of an attractive two-part tariff. The scheme was expected to become self-sufficient within a period of about five years.[53] In terms of scale, environment and density of population, the Norwich demonstration area was similar to the Bedford scheme. Centred on Reepham, about 12 miles to the north-west of Norwich, it covered approximately 125 square miles and sought to provide electricity for about 14,000 people living in just under 4000 dwellings. The cost of 60 miles of three-phase and 40 miles of single-phase transmission line was estimated to be £60,000 and the demonstration area was expected to be in surplus within seven years. In the interim, it would have access to borrowing rights from Norwich Corporation, while the Corporation would receive the financial backing of the Ministry of Labour.[54]

By the mid-1930s, then, a handful of government-backed schemes, together with the activities of a minority of forward-looking supply companies were beginning to spread the electrical gospel to the British countryside.[55] On a minority of arable farms, activities such as pumping, corn grinding, chaff cutting, threshing, sawing and sheep shearing were now more likely to be electrified.[56] In and around farmhouses and farm buildings electric light was no longer wholly novel.[57] In agriculture, as in industry, it was 'new' activities and sectors which made the most intensive and effective use of electricity. As early as 1927, the EDA was advertising automatic time-switches to lengthen periods of light during the winter and thus increase egg production on poultry farms.[58] By 1930 larger numbers of farmers were turning to poultry as a means of escaping bankruptcy and, although *per capita* consumption of eggs remained low in Britain compared with America and Canada, the large-scale production of day-old chicks became increasingly profitable.[59] In 1932 a progressive poultry farmer would be using electricity to heat batteries and incubators, grade eggs and test them for disease, grind food and provide power for plucking machines.[60] It is hardly surprising that a member of the EAW reporting on an Association visit in 1931 to the largest and most 'automated' of this new generation of poultry farms – Thornbers' of Mytholmroyd in the West Riding, the producers of a million chicks a year – should have titled her reflections 'A Million Balls of Fluff'. 'Electricity is used here in so many ways', she enthused, 'apart from ordinary lighting and heating. It is used to keep the requisite temperature in incubating and rearing rooms, to test the fertility of the eggs, to light the hen houses before daylight comes, so increasing the laying power of the hens, and to warm their drinking water.'[61]

The ubiquity of the new source of energy on a handful of farms as revolutionary, in terms of scale and 'mass production', as Thornbers', was well captured in 1937 by F. E. Rowland, a veteran observer of the agricultural scene: 'A chicken may first see the light of day when it is hatched in an electric incubator, from which it may pass to an electric brooder. When it is fully grown, it may go into a laying-house lighted and heated electrically. Its food can be prepared with electrically-driven machinery, and when its life is ended it may be plucked electrically, and finally be placed in an electric oven.'[62] In 1938, in typically cloying style, an anonymous contributor to the *Electrical Age* grappled with the problems of mass production and morality presented by the new 'chick factories'. 'A hen speaks!', she wrote. 'This light at night in the winter question. Personally I found it *very* misleading, oh, but very misleading indeed. Was it the sun, I asked myself, or wasn't it? And if it wasn't, then why did I find myself laying extra eggs? Speeding up production, that's what it is, and is it right, I ask you? Are we being exploited, girls? Should we form a union?'[63]

If the production of millions of day-old chicks would have been impossible without electricity, a more traditional sector – dairy farming – adapted more slowly. By 1933, though, modifications to the electric milking-machine, which made the sucking process approximate more closely to the movements of a calf's lips, were beginning to erode resistance to mechanisation.[64] Once milking-machines had been introduced – with estimated savings of up to 30 per cent – other tasks, on a minority of larger farms, began to be electrified. Automatic udder-clippers reduced the amount of hair finding its way into pails; cooling, sterilisation and storage systems were more frequently purchased; and pumping machines enhanced levels of hygiene in milking sheds.[65]

The mechanisation of traditional and innovative agricultural activities was matched, particularly after 1928 and the introduction of the National Mark Scheme, by the movement towards product standardisation. 'The key-note of the National Mark movement', it was stated in 1933, 'is standardisation – of product, pack and package. This standardisation leads to a greater degree of mechanisation, involving increased use of electric power.'[66]

By the late 1930s conveyor belts, refrigeration systems and temperature-controlled storage buildings, all dependent on electricity, were being used for the grading and marketing of soft fruit, dairy produce, poultry and eggs.[67] 'The very existence of the factory', Elsie Elmitt Edwards noted in 1936, 'is yet another proof of the value of electricity in rural areas. Instead of sending fruit to a town, losing its freshness en route, here [in Sussex] there is an estate, three hundred acres of which are already down to fruit, where the products can be dealt with on the day they are picked.'[68] By 1938 large-scale refrigeration was also playing a role in the development of the embryonic British fruit-juice industry.[69] Each of these developments was applauded by the now-depleted band of 'classical' electroculturalists. All represented the mechanisation of new and existing productive processes rather than the direct and 'revolutionary' application of electricity to arable crops vegetables and flowers.

Electrical innovation in agriculture, and forms of mechanisation which eased the introduction of urban styles of production to rural and suburban areas remained the exception rather than the rule in inter-war Britain. To the great majority of farmers, faced by domestic recession and the collapse of world markets, rigorous agricultural mechanisation meant next to nothing. It was against this overwhelmingly conservative background that triumphalists came to stress 'natural' connections between farming, the revival of the 'organic' village community, and the new form of energy as a stimulant to rural crafts and industries. 'We are in the midst of one of those crises in farming', an EDA pamphlet proclaimed in 1927, '[in which] it would seem that the present hope of relief is an ample supply of power'.[70] Electricity, Borlase Matthews insisted in the same year, held the key during a period which was likely to settle the 'fate of farmers' once and for all.[71] 'The ultimate wealth of any community', the same author insisted, 'is entirely dependent upon its success in agriculture'.[72] This neo-physiocratic mode of thought might seem to run counter to the technocratic and scientific mainstream of triumphalist and technocratic ideology, but it was widely subscribed to by planners, engineers and parliamentarians. 'It appears to be forgotten by all and sundry', lamented W. Fennell, an expert on wayleave law and a prolific writer, in 1929, 'that *the only way to preserve England is to preserve agriculture*'.[73] 'If we study history', Sir Douglas Newton told Parliament in the same year, 'we find that no country, however great, has retained its power and place in the world when it has neglected the interests of its rural community'.[74]

This preservation of the 'essential England' could be achieved, electrical progressives argued, via a rapid and concerted application of the new source of energy to farm and village. 'It would be unduly optimistic', *The Electrician* warned in 1930, 'to imagine that a large extension of rural electrification will, in itself, suffice to bring back prosperity to British farming. Yet it can render powerful aid in that direction, given an adequately co-ordinated system of low tariffs, assisted wiring, hire or hire purchase plans for the acquisition of machinery and a well considered instructional campaign.'[75] But, if farmers were to derive large-scale benefits, they must eschew conservatism and commit themselves to the ideals of the new scientific age. 'Should the farmers fail to realise the need for pulling together', *The Electrician* commented in 1931, 'in order to reap the benefits which the grid will bring within reach, the rural communities will have themselves to thank if they sink into a still worse state of depression than that from which they have been suffering for some time past.'[76] That some farmers were willing to take up the electrical standard was substantiated by the engineer-author of an article written in 1932 and entitled 'A Young Man's Outlook on the Electrical Industry'. 'Those [farmers] who are progressive and wise', he wrote, 'have during the last year or two realised that if electricity will not entirely solve all their troubles, it will go a long way towards giving them a decent chance of keeping their end up and eventually returning to a reasonable state of prosperity.'[77]

Advocates of the new source of power insisted that technical innovation and sustained cost-cutting could only be maximised if rural electrification could halt and in due course reverse 'the flight from the land'. In this idealised world, in which the 'tide of the rural exodus' would be stemmed,[78] it was electricity which would ease the introduction of new rural crafts and industries and revive old and ailing ones. Productivity and wages would rise; and the cultural and 'amenity' gap which had for so long separated town from country would be narrowed. Although writers and propagandists – many of whom were members of the Rural Reconstruction Association or the Rural Industries Bureau – advocated and gave total credence to the emergence of this new techno-arcadian order, they found it easier to point to foreign than indigenous examples of mechanically rejuvenated 'craft' and 'skill'. 'Wood-working and cabinet-making, weaving and pottery making' were unconvincing candidates for revival in the Britain of the late 1920s.[79] Sawmills and boat-making might well benefit from electric power, but 'hand-weaving, toy-making ... and leather-work' were more figments of the rural revivalist imagination than components of the real economy of the early 1930s.[80] And how many converts were likely to be recruited by the example of a 'clergyman ... making wooden angels with the help of an electric motor'?[81]

Borlase Matthew's accounts of the electrification of silk-weaving in Lyons, the mechanisation of brier-pipe production in the Jura mountains and the modernisation of spoon and fork manufacture in Savoy, seemed to belong to a backward and alien historical epoch.[82] And his assertion that in France 'the introduction of electric power into rural districts has caused many small industries to spring up like magic' smacked more of wish-fulfilment than controlled observation.[83] A tiny 'craft' sector in rural

Britain was unrealistically assumed to be ripe for electrification and mechanisation; and, if sceptics argued that modernisation, and higher productivity would lead to even higher levels of rural unemployment, electrical propagandists responded that the new form of energy would 'provide occupation for more hands' via increased regional economic activity.[84] Electric power would make possible 'to the enterprising craftsman all kinds of new developments by the introduction of new equipment, and if he seizes on the opportunity to introduce new lines of work he will find that, although electricity is replacing manual labour, it will at the same time provide occupation' for more men.[85] In this unreal world, in which 'mechanised craft' would supply rural and urban markets, rather than itself being crushed by large-scale town-based producers and retailers, village artisans would be transformed into electric mechanics. 'The smith of the future', insisted an editorial in *Rural Electrification* in 1930, 'will have to be a highly skilled all-round engineer, whose smithy will be equipped with up-to-date plant driven by electric power, and he will possess the skill equivalent to the skill of craftsmanship in days gone by'.[86] Thus it was that the needs of the new electrical age and of a buoyant rural community were reconciled: neo-physiocracy merged with triumphalism.

The rural reconstruction lobby was also convinced that urban areas, stricken by industrial decline and environmental impoverishment, could themselves derive salvation from the operation of the same 'natural laws' which were believed to underpin future agricultural recovery. 'When supplies of electricity were available in the countryside', Borlase Matthews told the annual general meeting of the Rural Reconstruction Association in 1931, 'many urban industries would migrate back to the healthier and more normal conditions of the country'.[87] There would be a 'return from the towns to the countryside, and the establishment of rural industries on a greater scale than ever'.[88] By now the revivalist blue-print, a product of intense anxiety over what seemed to be imminent catastrophe for British agriculture, had been expanded to incorporate deeply embedded cultural assumptions about the relative economic and moral worth of town and country.

A reborn agricultural sector would pull in labour from the socially bankrupt towns. This would be a gradual process – no reconstructionist foresaw empty towns surrounded by massively repopulated villages – in which commuters, bearing 'progressive' values from town to country, would play a crucial cultural role.[89] There was nothing unhealthy or socially debilitating about the fruits of scientific advance – every village would have access to a cinema, and every home would have a radio. The techno-arcadian dream reconciled a limited number of 'passive' urban spectator-activities with the dynamic, participatory and 'humanised' economy of villages in which 'craft' would be revolutionised and made more widely available. But the vices of the town would be left behind – the alienation, 'vulgarity' and 'unwholesomeness' of urban areas which had been shaped during the first, 'dirty' (and non-electrical) industrial revolution.

In reality, there was, of course, no 'mechanised craft revolution' in Britain in the 1930s; the idea that the social and economic existence of the 'smith, wheelwright, cobbler, market gardener, cucumber cultivator and bee-keeper' had been or could be transformed was vastly overstated.[90] But the vivid rhetoric and imagery of the techno-arcadian variant of triumphalism generated a literature – poems, short stories, novels, essays, newspaper editorials and 'countrymen's notes' – which reveals a great deal about the social history of inter-war Britain, particularly when juxtaposed against the militantly anti-technological modes of thought to be discussed in Part II. In 'Back to the Land', which was originally published in the *Observer* in 1933, the themes of urban decline, rural regeneration and scientistic progress were merged in a style simultaneously traditional and 'heroic':

> Sombre against the fading sky
> The pylons take the hill's long line
> Sombre . . . like monsters who divine
> Themselves held up to obloquy
> Lofty, unyielding, as they stride,
> Knowing what task they have in
>   hand –
>
> To lead men workless, wan, unmanned
> From slums to a greener country-
>   side . . .
> Yet not the same their sires did see,
> Where days with brutish toil were
>   filled;
> To hamlets lighted now – land tilled
> By harnessed electricity![91]

In this workaday poem interactions and contradictions between arcadian, techno-arcadian and reconstructionist ideologies and mythologies are revealed – social resistance to the new 'age of the pylon'; the remorseless and 'sombre' progress of a massive technological network; an unchanging and restorative rural 'wilderness', unostentatiously mechanised by the new source of power; and a debilitated and 'unmanned' urban sector, crying out for the healing benefits of the 'lighted' hamlet. Taken together, these constitute a world-view which seems to square the cultural circle. The ravages of industrialism and the 'machine' were to be ameliorated in a 'clean' and 'communal' agrarian environment, itself renovated and revitalised by the new source of energy. In reality, of course, the land had not been electrically 'harnessed' in Britain in the 1930s. Nor had technocrats, planners of the National Grid, or rural supply engineers succeeded in abolishing 'brutish toil'.

## Notes

1   Borlase Matthews's life is under-documented, but see the obituary notices in *Journal of the Institution of Electrical Engineers*, 90 (1), 25, 1943, 539, and *The Engineer*, 27 August 1943, 159.

2   *ER*, 15 December 1922, 891. Similar complaints were made in the same journal throughout the early 1920s. See 13 April 1923, 563; 29 February 1924, 323; and 14 November 1924, 547.

3   *ER*, 12 December 1924, 883.

4   *Hansard*, 21 May 1925, 741.

5   *ER*, 12 March 1926, 406.

6   F. Ringwald, 'Electricity in Agriculture' *E*, 17 September 1926, 322–3.

7   *EII*, 9 October 1929, 1657.

8   *Hansard*, 232, 2 December 1929, 1959.

9   Ibid 238, 12 May 1930, 1519.

10  *RE-EF*, April 1932. For an equally pessimistic assessment at this time see *Hansard*, 273, 21 December 1932, 1064. Sir W. Wayland, member for Canterbury.

11  H. J. Denham, 'The Needs of the Farmer and the Responsibility of the Supply Undertakings', *PIMEA*, 1933, 228–9.

12  F. E. Rowland, 'Country and Farm', *E*, 27 April 1934, 557.

13  R. Borlase Matthews, 'Rural Electrification', *E*, 25 January 1935, 111.

14  Lord Eltisley, 'Rural Electrical Development', *ER*, 12 June 1936, 865–6. For further comment on the perceived 'distraction' of electro-culture see F. E. Rowland *op cit*, 558.

15  On this contentious issue see *Hansard*, 314, 14 July 1936, 1997, J. P. Maclay, member for Paisley; *Hansard*, 321, 8 March 1937, 803–4, Sir Percy Hurd, member for Devizes; and Caroline Haslett, private evidence to the McGowan Committee, 37. But note, also, the counter-blast decrying the unreasonable expense of connecting up rural consumers in *E*, 24 February 1939, 226.

16  *Hansard*, 335, 27 April 1938, 117.

17  'District Engineer', 'Electricity in the Countryside', *E*, 31 March 1939, 397–8. Similar points were made by 'An Electrical Contractor' in 'Electricity in the Countryside', *E*, 10 March 1939, 305.

18  *Hansard*, 346, 26 April 1939, 1134–5.

19  'The Clyde Valley Electrical Company: Quarter of a Century's Work',

*EAW*, April 1931, 139–40. See, also, 'Rural Electricity Supply in the Clyde Valley Power Co.'s Area', *EF*, March 1928, 303.

20 'Electrification of Farms in Fife', *RE–EF*, February 1937, 199.

21 J. S. Pickles, 'Rural Electrification', *E*, 10 December 1937, 695–6 and *idem*, 'Rural Electrification: the Development of the Dumfries Rural Area', *RE-EF*, January 1938, 127–133.

22 See, on this issue, *Hansard*, 318, 24 November 1936, 242, A. G. Erskine Hill, member for Edinburgh North; 326, 7 July 1937, 339–40, Malcolm MacMillan, member for the Western Isles; and 347, 10 May 1939, 480–1, W. Gallagher, member for West Fife.

23 *Hansard*, 233, 11 December 1929, 493, D. Hopkin, member for Carmarthen.

24 Ibid, 511. On the refusal of Welsh companies to connect up householders whose tenancy appeared to be under threat see Caroline Haslett, evidence to the McGowan Committee, 20.

25 'The Cowbridge Rural Area', *RE-EF*, January 1937, 168. See also the earlier survey in W. A. Chamen, 'Development of Electricity Supply in Rural Districts in South Wales', *EAW*, July 1927, 184.

26 W. H. Jones, 'Development in Uses of Electricity in Rural Wales', EDA 1475 (1938), 1–6.

27 Laurence J. Kettle, 'Electricity Supply in Ireland', *E*, 5 November 1926, 533. See, also, J. Milne Barbour, *RE-EF*, December 1938, 139 and Michael J. Shiel, *The Quiet Revolution: the Electrification of Rural Ireland 1946–1976* (Dublin, 1984).

28 See *E*, 7 January 1921, 44, and 20 July 1923, 53 and *ET*, 11 March 1926, 308.

29 *ET*, 27 March 1924, 354.

30 R. Borlase Matthews, 'Electro-Farming: a New Type of Tractor for Electric Ploughing', *ER*, 2 July 1926, 5. But note the comments on the impracticality of electric ploughing in J. G. B. Sands, 'Electricity in Agriculture', *E*, 27 September 1929, 370–1.

31 *E*, 9 July 1926, 52.

32 *EF*, June 1927, 5.

33 Sven Oden, 'Electro-Culture of Plants', *ET*, 15 May 1930, 965.

34 'Report to the Electricity Commissioners' Rural Conference on Electricity in Agriculture and Horticulture', EDA 1042 (1932), 26.

35 Sir William Ray, 'The Application of Electricity to Agriculture', *EF*, May 1934, 374.

36 Ibid.

37 *ER*, 14 November 1924, 722.

38 'Will Electricity Save England?', *ET*, 3 December 1925, 669.

39 *ER*, 25 June 1926, 943.

40 'Electricity Supply in Rural Areas', EDA 681 (1927), 6.

41 *ET*, 20 October 1927, 482. A pressure group – the Overhead Association – committed to reducing the red-tape alleged to have inhibited electrification was founded in 1927. See *ER*, 21 October 1927, 659.

42 *EF*, June 1928, 6.

43 *RE-EF*, February 1929, 262.

44 Margaret Partridge, 'Wayleaves', *EAW*, January 1930, 572.

45  See Hannah, *Electricity Before Nationalisation*, 116.
46  'Supply Engineer', *ER*, 8 November 1935, 633.
47  *E*, 16 December 1938, 715.
48  The full case against the companies is eloquently made by Margaret M. Partridge, 'How Electricity is Supplied to Rural Districts', *EAW*, April 1928, 301–2; Caroline Haslett, evidence to the McGowan Committee, 22–3; and Electricity Commission, *Report of Proceedings of Conference on Electricity Supply in Rural Areas* (HMSO, 1928).
49  *E*, 17 June 1921, 742 and *E*, 29 September 1922, 339.
50  S. E. Britton, 'Five Years' Progress in the Electrification of Agriculture around Chester', EDA 1179 (1933). See, also, the same author's 'Rural Electrification Experience', *ER*, 14 July 1933, 49; and 'Chester – a Record of Intensive Rural Electrification', *RE-EF*, November 1935, 171. Aylesbury had pioneered a rural scheme from the mid-1920s. (*EAW*, October 1926, 49.)
51  S. E. Britton, 'Rural Electrification Experience', *ER*, 14 July 1933, 49.
52  *EF*, December 1927, 211. See, also, *E*, 18 November 1927, 641.
53  'Rural Electrification', *E*, 21 February 1930, 242.
54  'Rural Electrification: a Memorandum on the Norwich Demonstration Scheme', *E*, 27 March 1931, 491.
55  For evidence of such developments see 'Norwich – Further Progress in Rural Electrification', *RE-EF*, August 1935, 140 and R. Borlase Matthews, 'Electro-Farming', *E*, 28 January 1938, 108.
56  'Electricity and the Farm', EDA 1045 (1932).
57  F. E. Rowland, 'How Electricity Helps the Farmer', EDA 1044 (1932), 3–7.
58  'More Eggs in Winter', EDA 670 (1927).
59  'Electricity in Poultry Farming', EDA 874 (1930).
60  'Electricity on the Poultry Farm', EDA 981 (1932).
61  Eleanor G. Johnson, 'A Million Balls of Fluff', *EAW*, October 1931, 236. See, also, Anna Holm, 'Discover What Electricity is Doing for the Rural Dweller', *EA*, July 1932, 384, for further detail on the 'poultry revolution'.
62  F. E. Rowland, 'A New Aid for the Farmer', *EA*, April 1937, 213.
63  'Harriett the Hen!', *EA*, Summer 1938, 437.
64  Eric Brain, 'The New Farming Age', EDA 1093 (1933).
65  'Electricity and Lower Milk Production Costs', EDA 1301 (1935); see, also, H. J. Denham, *PIMEA* (1933), 231.
68  Elsie Elmitt Edwards, 'Fruit Canning in Sussex', *EA*, October 1936, 138.
69  'Ceres', 'The Health of a Nation', *RE-EF*, November 1938, 102.
70  'How To Make Farming Pay!!', EDA 672 (1927).
71  *ET*, 26 May 1927, 712–13.
72  R. Borlase Matthews, *Electro-Farming or the Application of Electricity to Agriculture* (1928), 2.
73  W. Fennell, 'Rural Electrical Development', *ER*, 18 October 1929, 646.
74  *Hansard*, 233, 11 December 1929, 550.
75  'Rural Electrification Policy', *E*, 30 May 1930, 665.
76  'Rural Electrification', *E*, 24 April 1931, 609.
77  R. C. Hawkins, 'A Young Man's Outlook on the Electrical Industry', *EII*, 27 June 1932, 1169.

78   Lady Snell, 'Electricity and Village Industries', *E*, 25 September 1929, 350.
79   H. F. G. Woods, 'The Distribution of Electricity in Rural Districts', *PIMEA*, 1928, 81.
80   Alan Thwaites, 'Rural Electrification in Yorkshire', *EAW*, April 1931, 146.
81   'Harriett Holmes' (Caroline Haslett), 'A Success in Rural Electrification', *EA*, October 1932, 399.
82   Borlase Matthews, *Electro-Farming or the Application of Electricity to Agriculture* (1928), 113.
83   Ibid, 114.
84   *EII*, 18 September 1929, 1531.
85   *RE-EF*, October 1929, 133–4. This was a quotation from a BBC booklet.
86   *RE-EF*, January 1930, 230.
87   *ER*, 10 April 1931, 647.
88   *RE-EF*, December 1931, 202.
89   *ER*, 29 April 1932, 635.
90   Vera Norwick, 'Rural Industries and Market Gardening', *EA*, July 1932, 349.
91   H. Baines, 'Back to the Land', *EA*, January 1933. Originally published in the *Observer*.

# Part II

Resistance

**Figure 5** The National Grid as planned, *c.* 1930. Source: *Annual Reports* of the Central Electricity Board

# 6

# Downs, lakes and forests

For electrical progressives, the 4000-mile National Grid, which had been given legislative blessing by Baldwin's Conservative government in 1926, represented the imminent supremacy of the new form of energy. Triumphalists believed that the construction of a national network would significantly reduce unemployment and provide a much-needed impetus for the national economy. Electricity, in its domestic role, would deliver heat and light to benighted urban areas and encourage the introduction of new forms of industry to regions decimated by depression. This was the dream, and it has been described in detail in Part I. But a minority of triumphalists acknowledged that in the short term indifference and hostility might well be encountered. The attractiveness of electricity would eventually become self-evident, but in the interim, large-scale technological innovation would transform the face of the nation without bringing rewards to those most directly affected by the presence of new and alien physical structures. The natural landscape and 'amenity'; farming land and property rights – all would be threatened by the 'march of the pylons' and the building of massive and polluting 'super-stations'. Yet progressives remained convinced that the 'rationalisation' of 'national electrical development' was all-important. Compared with other European nations, in which the state had adopted a strongly interventionist role, Britain continued to be dogged by inefficiency. About thirty substantial supply companies now claimed a lion's share of the national market but the existence of approximately five hundred much smaller concerns ensured that there continued to be large cost and price differentials. Widely differing policies over direct and alternating current, domestic voltage levels, wiring schemes and tariff scales ensured that Britain constituted not one, but many 'electrical nations'. The Grid would co-ordinate the production and interconnection (though not the distribution) of national supplies, force small and uneconomic concerns out of business, and cut the price of electricity for every category of consumer. Triumphalists were certain that the 'community' must be willing to pay high short-term costs to achieve so majestic a scientific and cultural transformation. But others remained unconvinced. By far the most threatening of these 'anti-electrical' elements were the protest groups which established themselves in the late 1920s and early 1930s in the South Downs, the Lake District, the New Forest, and London. In Scotland, also, in the mid-1930s preservationists tried to prevent the British Oxygen Company from developing Highland loch water for the production of hydroelectricity. It is to the organisation, activities and motivation of each of these movements that we now turn.

The South of England scheme, affecting forty-seven miles of magnificent Sussex countryside, was published by the Electricity Commissioners in September 1927, and adopted by the Central Electricity Board in February 1928. The route was surveyed in June of that year and bargaining for rights of way began immediately.[1] This involved negotiations both with individual landowners and local authorities; and the Board became adept at persuading both groups that the national interest demanded that the building of the network must not be delayed.[2] When these techniques failed, as they did in rural Sussex in 1929, the Minister of Transport was empowered, though not compelled, to give local authorities and other interested parties an opportunity to be heard at a public inquiry.[3] The South Downs inquiry was held at the Town Hall, Eastbourne on 11 and 12 September 1929. The Ministry of Transport inspector was disconcerted to find himself opposed by large numbers of articulate landowners and by environmental pressure groups – the Council for the Preservation of Rural England, the Sussex Downsmen, and the Sussex Archaeological Society. Several local authorities were also represented, the most militant of which proved to be the East Sussex County Council. J. E. Seager, the deputy clerk, began the attack by claiming that no investigation had been undertaken into the possibility of placing the lines underground rather than overhead and that, if the scheme went ahead in an unmodified form, the social costs of rural electrification would far outweigh the gains.[4] Reordering the history and prehistory of the Sussex landscape, Seager argued that 'nature has preserved the bare simplicity of the Downs for thousands of years. Our countryside is our life-blood and our heritage. We have in the South Downs something unique and something which only nature can give. The Downs should not be ruined by a short-sighted scheme when in a few years science may find some means of laying these cables underground.'[5]

The Sussex Archaeological Society contended that the erection of the pylons would destroy numerous ancient sites and monuments, while, from a quite different perspective, Lord Gage, for the Central Landowners' Association, deplored the long-term impact on land values.[6] Others went further: the entire 'agricultural character' of the region was under threat.[7] Opponents of the scheme were also scathing about the way in which the inquiry had been conducted. The technical issues which underlay the dispute, and the unavailability of data to whose who opposed the scheme, put the protesters at a profound disadvantage. This point was most succinctly expressed after the inquiry by Patrick Abercrombie. 'Is there any means', he asked, 'of reviewing the national scheme as a whole, not leaving individual gangs of Downsmen or Lakedwellers to fight an unequal battle against electrical mandarins who floor them with technicalities often contradictory and often disputable?'[8]

In the aftermath of the inquiry, before the Minister of Transport, Herbert Morrison, had announced his decision, opposition continued to mount both locally and nationally; and, by late September, Morrison was rumoured to be under heavy pressure to modify the scheme.[9] Eminent writers and intellectuals, including Keynes, Galsworthy, Kipling and Belloc, who either lived in or had close connections with the Downs now lent their authority to the anti-pylon movement.[10] Protection of the environment and what was becoming more widely known as 'amenity' dominated these initiatives. 'When we contemplate the fair face of the Sussex Downs', protested Colonel R. V. Gwynne, the Mayor of Eastbourne, 'scarred with masts and cables, our picturesque villages imprisoned in a wire cage, and the placid beauty of the Pevensey marshes slashed with a line of steel towers, we think that this is the type of progress that we could very well be without.'[11]

When, in mid-October, Morrison announced that, with a single minor modification, the scheme would go ahead, the protesters' hopes seemed to have been demolished. Yet it soon became apparent that this decision had united and strengthened local and national opposition. The environmental case against the pylons was comprehensively restated and there were angry denials that the completion of the Grid would lead to increased rural electrification. As for Morrison's suggestion that the scheme could be reviewed after five years – this was dismissed as 'ludicrous' (as indeed it probably was).[12] In terms of amenity it was said that the state preached one thing to its citizens while practising another. Awareness of the need for protection of the environment was now stronger than ever, and on all sides battle was being waged against

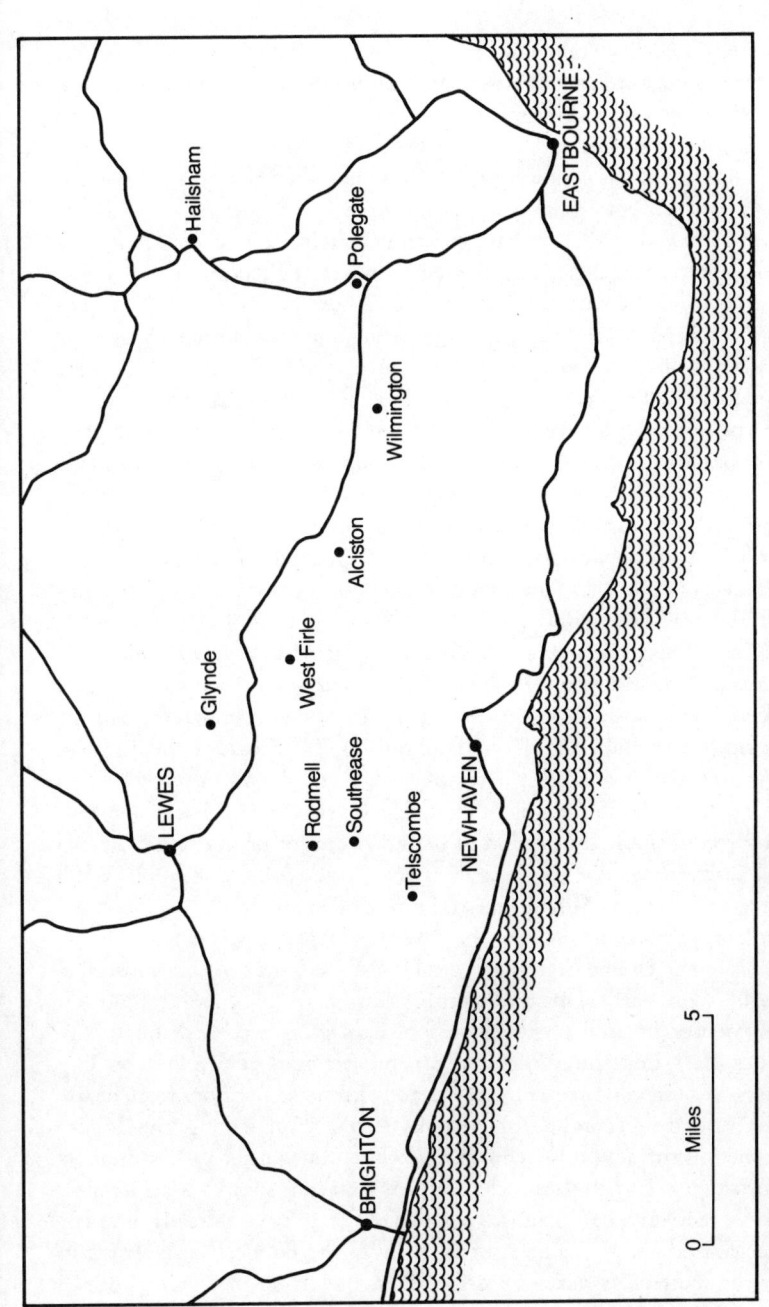

**Figure 6** Centres of electrical debate in the South Downs and adjacent areas, summer to winter, 1929

'uncontrolled development, inharmonious buildings, overhead electric transmission, outdoor advertisements, ill-designed petrol-filling stations, refuse dumps and indiscriminate tree-felling'. But how was this work to be carried on if ministers of the crown were willing to ride roughshod over informed local opinion?[13] By this juncture, however, a national counter-attack was gaining momentum. 'Stick it, Herbert', urged the *Daily Herald*, 'even if you have to paint them [the pylons] green'.[14] The artist–intellectual, Eric Gill, pointed to the contradictions of the 'purist' conservationist case. 'An attachment to nature', he urged, 'which goes with a refusal to see beauty in engineering and making money by it, is fundamentally sentimental and romantic and hypocritical. Let the modern world abandon such an attachment or let it abandon its use of electric power.'[15] Sir Reginald Blomfield, who was employed on a part-time basis by the CEB as a consultant on landscaping, believed that 'anyone who has seen these strange masts and lines striding across the country, ignoring all obstacles in their strenuous march, can realise without a great effort of imagination that [they] have an element of romance of their own. The wise man does not tilt at windmills – one may not like it, but the world moves on.'[16]

The technical press was even more outspoken. Insecure and buffeted by recession, the industry was convinced that *agents provocateurs* were fomenting opposition to the pylons and that such manoeuvres had the backing of gas and coal.[17] Even more sinister, *The Times* and the BBC had now joined the anti-electricity lobby.[18] 'Artists' and 'aesthetes' who placed 'obstacles in the way of national efficiency' must be uncompromisingly combated. Opposition to the new form of power was irrational and reactionary. But the leaders of the Sussex protest movement still dared to hope that Morrison – and if not Morrison, then MacDonald – might be persuaded to modify the Downland scheme. MacDonald was on a visit to the United States, and a cable had been sent to the embassy in Washington earlier in the month. 'Knowing of your feelings on preservation of rural amenities we appeal to you to receive deputation re South East England Electricity Scheme on your return.'[19] Downing Street was then bombarded with carefully prepared petitions from nearly forty rural district and parish councils, strongly supporting 'the County Council in the action which it has taken' and urging it to 'continue its efforts to preserve the beauties of Sussex from spoilation'.[20] MacDonald's private secretary, C. P. Duff, argued that the Premier was already seriously overworked and that a delegation ought not to be received.[21] Morrison was more direct. It was not possible, he wrote, 'to suspend work with all the consequences of delay upon our electrical development and upon the employment situation during a period of further inquiries and investigations'. He also insisted that 'from the point of view of amenity alone the agitation is misconceived . . . the interference with the landscape will be infinitely less than is imagined or is represented by the grossly misleading composite pictures which appeared in *The Times*'.[22] But Morrison was overruled and a deputation, led by Earl Buxton, an active conservationist, was received by MacDonald and his minister on 25 November. Morrison was closely questioned by J. E. Seager, by Colonel A. S. Sutherland-Harris, chairman of the East Sussex County Council, and by Buxton. But he was well able to look after himself. Morrison kept to a rigidly legalistic position – as minister, he had no option but to carry out his duties under the electricity Acts. It was technically and socially impossible for rural Sussex to opt out of the national scheme. To change the Downland route – to take the pylons across the Weald, as had been suggested on a number of occasions by the protesters – would be too expensive and lead to objections from those who believed the Weald to be just as beautiful as the Downs. Where Morrison was terse and unapologetic, MacDonald, who had gained something of a reputation as a lover of the countryside, was more conciliatory.[23] Now he found himself having to square the circle. Rural preservationism was a vital issue, but Sussex could not be relieved of its duty to participate in the construction of a national electricity network.[24]

By late 1929 there were strong feelings of bitterness and defeat in rural Sussex. The final insult, or so it was interpreted by the anti-pylon protesters was a well-authenticated rumour that a last-minute delegation from Mayfield Parish Council to the Board had been informed, 'we don't care if all the people in Sussex protest: it is going forward'. 'The old idea', lamented the anti-centralist *Sussex Chronicle*, 'that the English system of government was . . . of the people, by the people, for the people only survives as a battered tradition but it is seldom that the new form of government of the people by autocratic officials is proclaimed with such brutal frankness.'[25] But there were compensations, albeit minor ones. MacDonald was more sensitive to public opinion, whether articulated by artistocrats or the labour movement, than the arch-realist Morrison. There are indications that the Prime Minister believed that the Downland pressure groups had been undiplomatically treated by his Minister of Transport.[26] As early as August, MacDonald had received a delegation from the CPRE to discuss ways in which the government might sponsor and legislate for the protection of the environment. The memorandum which the Council presented emphasised that main roads should be more carefully planned and that unrestricted development around and alongside them more strictly controlled. The government was also urged to designate national parks; to force recalcitrant local authorities to make use of existing planning provisions; and to attend to aesthetic aspects of large-scale housing projects. More money, it was argued, should be spent, through the Board of Education on 'civic subjects' to increase young people's awareness of rural matters. In a section dealing with electricity, it was stated that 'much perplexity exists in the public mind in connection with the ... schemes, not only with regard to the relationship between the various departments and interests concerned, but also with regard to procedure'.[27]

At the very end of his career Morrison would characterise the CPRE as a group of well-meaning but wrong-headed zealots who had a minor role to play in the corporate, rather than participatory democracy of which he was so staunch an advocate: they were little more than a useful spur to 'efficient administration'.[28] It would be foolish to allow such people too close to the levers of power. The Council had a quite different impression of things, and considered that a landscaping circular which was sent to all electricity supply authorities in April 1930 represented a major advance.[29]

The collapse of the South Downs movement appeared – wrongly, as it transpired – to mark unqualified defeat for the anti-pylon campaign. Certainly, this was how it was perceived by the ever-watchful electrical press. 'The case', the *Electrical Review* observed, 'will serve as a precedent in other districts, which cannot pretend to the same eminence as the South Downs, and where legitimate objections can be met by minor modifications which will not appreciably interfere with the general scheme.'[30] Sussex had strong historical connections with London and central government. It was dominated by powerful landowners and articulate intellectuals. If opposition could be outmanouevred here, it could be outmanouevred anywhere.

Attitudes towards pylons, particularly among the intelligentsia and that much-derided group, the 'aesthetes', began to change during the early 1930s. 'The Pylons are coming', declared the *London Evening News*, 'the champions of aestheticism – even the Champions of Unspoiled Rural England – are not unanimous in thinking that pylons *are* repulsive. Clearly their difficulty in agreeing is civilisation's opportunity.'[31] What may have been either an acknowledgement of defeat or an adjustment to the inevitable was now strongly reflected in the CPRE's more moderate attitude towards the Grid as well as in professional and academic rationalisations of the interaction between the 'natural' and the 'mechanical'. In a seminal article, the *Architect's Journal* voiced this newly emerging orthodoxy. Only in very ancient times had Britain been blessed with a 'natural environment'. Each epoch had witnessed successive man-made accretions to the landscape, and each of these had redefined the visual impact of the environment as well as the meaning of 'amenity' itself. Electricity pylons had now begun to merge into this 'man-made nature', to join hedges, ditches, farm buildings, townscapes, factories and arterial roads.[32] Perhaps H. G. Wells's 'Martians riveted into a landscape' could be lived with after all?[33]

The South Downs anti-pylon movement had been defeated. But why? First, the protesters had organised themselves slowly and amateurishly. To give a single, glaring example, Newhaven Rural District Council had still been in the throes of deciding what its attitude towards the anti-electric deputation should be when the deputation had already returned from Downing Street.[34] Secondly, there was a pronounced cleavage of opinion between local authorities representing rural constituencies and those which championed urban and commercial interests. This can be demonstrated by an analysis of the petitions sent to MacDonald immediately before he received the final delegation. Brighton stood conspicuously and consistently aside from the conflict and may be said to have tacitly supported Morrison and the CEB.[35] Newhaven took up a more militantly anti-conservationist stance. In a chamber of commerce debate held in the immediate aftermath of the controversy, 'amenity' was defined not in its environmental sense but with reference to the 'domestic amenity' of cheap electricity; and Downland farmers were criticised for putting up barbed wire to keep working-class ramblers away from the most attractive parts of the countryside.[36] The protesters also failed to offer a convincing explanation of why the South Downs and not the Weald should remain sacrosanct. This indicated a deeper and more debilitating weakness. To many, 'amenity' seemed a vague and fundamentally relativistic concept. Might it not be little more than a slogan to protect the interests of a rural leisure class? The most convincing explanation of why the Sussex anti-pylon movement failed is that it was overpowered and outwitted by an increasingly powerful state bureaucracy. Deprived of vital technical data, denied an opportunity to debate the regional scheme within the context of the Grid as a whole, and only rarely able to engage on an equal footing in the scientific–judicial discourse which marked the CEB's presentation and legitimation of its actions – for all these reasons, those who resisted the pylons became enmeshed in a truly 'inauthentic' form of communication. Whatever may have been the 'objective' balance between pro- and anti-electric forces in the country as a whole, this growing and historically rooted imbalance between 'state centre' and 'periphery' was to prove decisive.[37]

As the South Downs saga moved towards its climax, anti-pylon activity was intensifying in the Lake District. The publication of the Central Electricity Board's scheme for the north-west section of the Grid in 1928 had concentrated the minds and energies of an already well-established regional environmental movement. As early as 1927 representatives of the Society for Saving the Natural Beauty of the Lake District (SSNBLD) had entered into informal discussions with Sir Andrew Duncan in London in the hope that it would be possible to persuade him that the beauty spot of Whinlatter Pass, near Keswick, should remain untouched by large-scale pylon contruction.[38] Now, in the summer of 1928, preservationists were also advocating the 'total omission of Keswick from the scheme' and insisting that as much cable as possible should be placed underground.[39] By early 1929 a group of interested individuals in Keswick, with strong connections with the CPRE and with metropolitan environmental and academic élites, was pressing ever more insistently to keep the Grid away from Whinlatter Pass.[40] But when Keswick Council held its first full debate on the issue in June of that year, the scales were delicately balanced, and the motion that the CEB be requested to place its cables underground inside the town itself was only carried by a single vote.[41] If there was any agreement at local level between those who opposed the National Grid and those who believed that cheap electricity would boost both rural and urban economic activity, it was that tourism was now central to the well-being of the Lakes. There was also consensus that the Lake District was in danger of being desecrated by every kind of random 'development', electrical and otherwise. 'Keswick', a local newspaper stated, 'depends for its livelihood on people who want to get away from the pylons and all that they stand for, and the time is not far distant when the district will be so shriven by road surveyors, so pilloried and posted by electricity and telegraph engineers, that it will not be worth coming to see.'[42] *Country Life* would soon view the issue in larger and more symbolic terms. 'The agitation in the Lake District is so important because it is a test case, the first trial of the legal claims of the Old England versus the New.'[43]

By the summer of 1929 the architect and preservationist, Kenneth Spence was leading a campaign to link opposition in Keswick to the

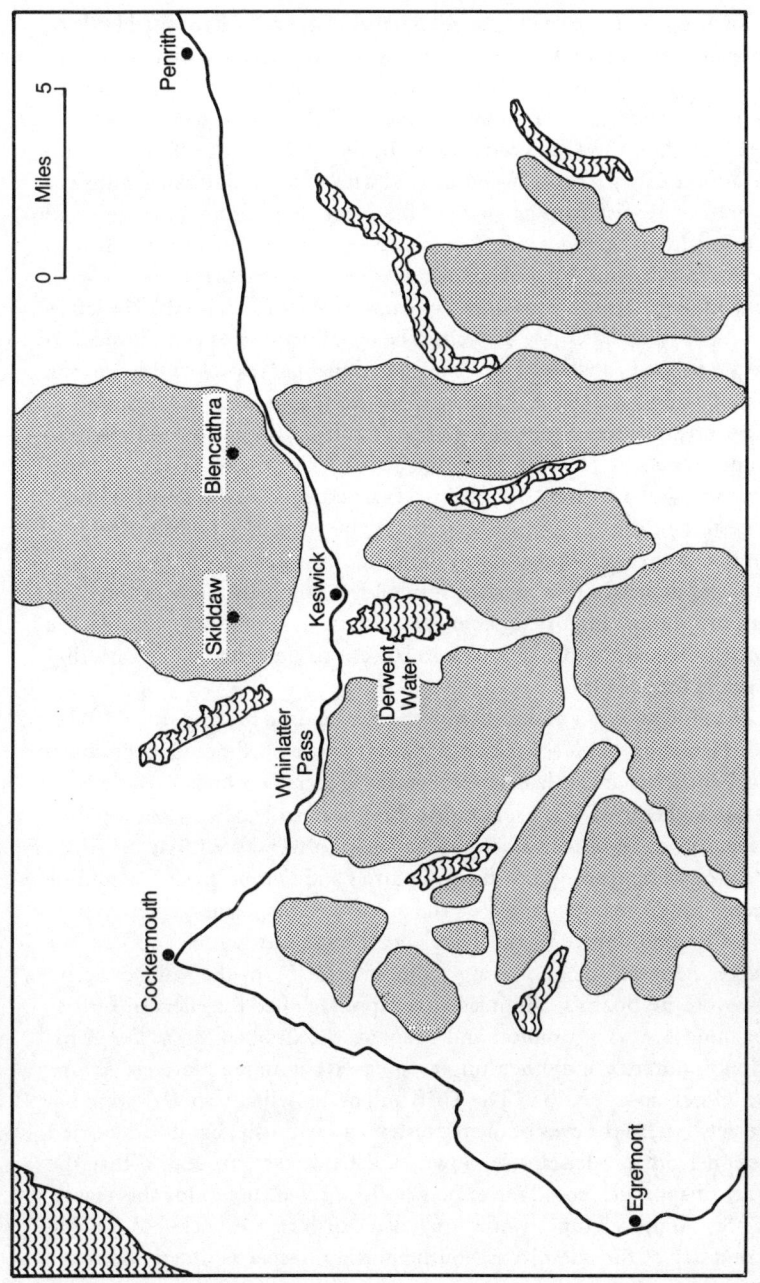

**Figure 7** The area at issue during the Keswick debate, 1929–33

national anti-pylon movement and to an already well-established body of environmental opinion which wished to see the Lake District converted into the first of Britain's national parks.[44] 'The whole of the Lake District', the *Cumberland and Westmorland Herald* insisted, 'ought to be a national park, bought up and reserved for ever by the state.'[45] Hugh Walpole, who had adopted the Lakes as his home, and who wrote and spoke in a variety of preservationist forums in the late 1920s and early 1930s, referred emotionally to the uncertain future of Skiddaw and Blencathra. 'It is to the interest of every English man and woman who cares for the beauty of this country to see that this small and famous plot of ground shall be left in its proper peace.'[46] G. M. Trevelyan believed that it was commitment to private property that was inhibiting a national parks policy: this was the context within which the threat to Keswick must be seen as a 'test case'.[47] Local pressure for a full-scale public inquiry now burgeoned but was rapidly rebuffed. Herbert Morrison had been appalled at the national coverage given to the 'anti-electrical' arguments of the South Downs protesters in the run-up to the public tribunal in Eastbourne, and was convinced that all opposition to the Grid must henceforth be 'administratively' defused. This meshed in with the dominant view at the CEB that the public inquiry system was costly, inherently hostile to the 'electrical cause', and likely to lead to serious delay to the engineering schedule for the Grid.

Thus it was that a month before the Eastbourne hearing, the CEB set about isolating the Keswick anti-pylon movement. Since Cockermouth and Penrith councils had agreed to the general outline of the north-western scheme, it was argued that there was no legal case for a public hearing in relation to Keswick. The Board also insisted that it had no intention of venturing into the town itself and that no pylons would be erected inside its boundaries. To the protesters this seemed a cynical ploy which evaded debate of the Lakeland scheme as a whole and drove a wedge between neighbouring communities.[48] In the longer term, however, the Board's 'administrative' approach had the effect of further alienating Keswick opinion and drawing the Council, an *ad hoc* Anti-Pylon Committee and other long-established community interest groups into closer co-operation. The CEB might be willing to act 'administratively', or, in the eyes of the protesters, imperiously; but it had learned enough from recalcitrant landowners and farmers, to realise that the informal approach could never be wholly rejected. It was for this reason that Sir Andrew Duncan now travelled north to Keswick and, on the second day of the fateful Eastbourne hearing, inspected the pylon route

*Downs, lakes and forests*

with John Bailey, chairman of the executive committee of the National Trust and Dr Anderson, a leading member of the Anti-Pylon Committee. The protesters doubted whether the conditions for negotiation with the CEB had been rigorously enough defined, and whether Bailey would drive a hard enough bargain. Initially, all seemed to go well. Bailey reported that Duncan was an exceptionally tough adversary. 'He has a public duty to perform, that of providing cheap and universal electricity, and he will not mind me saying that he is not a Scotsman for nothing.'[49]

No agreement was arrived at following the examination of the route, but Bailey was given informal authority to continue his discussions with Duncan in London. Within a matter of weeks these negotiations and the agreement which flowed from them, would be denounced as acts of treachery. The future of the Lakes, it would be claimed by Kenneth Spence and his anti-electric supporters, had been decided by Londoners with little knowledge of local opinion; the National Trust had shown itself to be too easily suborned by government; 'soft' southern environmentalists had been too ready to disown their 'unreasonable' northern colleagues. The root of the problem lay in the vagueness of the enabling motion passed by Keswick Council, but the Trust had failed to keep the Council informed of what had gone on behind closed doors in London. The agreement between Bailey and Duncan promised a small reduction in overhead cable. The pylons themselves could be painted any colour acceptable to local opinion and individually evaluated in terms of placement, with the National Trust acting as intermediary between the CEB and the community. There would be a small amount of underground line in Keswick itself, and a review of the total position after seven years. The Board had been pressed to move the whole scheme north of Skiddaw but had refused on the grounds that additional costs in the Lake District had already reached £30,000 and would have been doubled had such a concession been granted.[50]

But the Keswick anti-pylon movement was unimpressed with the CEB–National Trust accord and there was now, in October 1929, at the height of the identifiably national phase of the anti-electric campaign, a renewed determination to achieve absolute unanimity between Keswick Council, the Anti-Pylon Committee and other local organisations.[51] In a wide-ranging attack on the Duncan-Bailey agreement, Kenneth Spence pointed out that, in terms of an increase of underground cable in Keswick and its immediate environs, 'what the Board has offered us is two sections of about half a mile each instead of the $4\frac{1}{2}$ or $6\frac{1}{2}$ miles we consider essential'. He also doubted whether the existing private electricity company would capitalise on the presence of the Grid to provide Keswick with cheaper power. 'Penrith and Keswick', he concluded, 'are non-industrial towns with populations of about 8000 and 4000 respectively, both of which already have their own supplies of electricity.' The CEB was now talking in terms of the positive impact of the new system of electrical supply on agricultural as well as industrial development. But 'what agricultural development', Spence asked, 'can they expect in this sheep-grazing district?'[12] By early 1930 Keswick had been almost wholly isolated by the CEB. Construction between Penrith and Carlisle was continuing apace and information trickling westward about the gigantic lines of pylons. By mid-summer the CEB was entering into rapidly negotiated wayleave agreements with farmers on the very borders of the town; the recalcitrant community was being made to sit and sweat it out. Little wonder, then, that the eleven man Council was still wavering, as it had done from the very beginning, between 'no-pylon', 'buried cable' and 'cheap electricity' strategies.[13] But it was not until 1932, when national opposition to the Grid had significantly diminished, that the CEB decided to grasp the nettle for what it hoped would be the final time. On this occasion it was the Board's deputy secretary, O. A. Sherrard, who travelled to the North-West to meet the Council and the Anti-Pylon Committee. The discussions were tense. Sherrard refused to budge from the Duncan–Bailey accord and hinted that, *in extremis*, the Board still possessed the right to override all local opposition.[14] During a particularly stormy session Sherrard told a councillor, 'You want more than the National Trust, you are more Royalist than the King'. 'And I am proud to be', came the reply.[15] The Board was accused of surrounding the town with pylons and bullying landowners if they refused the wayleave rentals offered to them. 'The people who built the railway', another councillor informed Sherrard, 'had a jolly sight more sense than the people who planned the pylons, for they did put it in the valley out of sight, but you are going to put three pylons right on the skyline just where people come into Keswick and look down its valley. They will strike the eye at once and be an absolute eyesore.'[16]

But if the CEB came in for heavy criticism, so also did the National Trust, which was held to have fatally compromised the activists' position. Kenneth Spence now believed it essential to create a united environmentalist front but, when he approached the CPRE, H. G. Griffin, the general secretary, was still chary of offending the National Trust. 'Time is short', Spence retorted, 'and if you are a real CPRE you will not wait for formalities with the NT.'[17] In late January 1932, Keswick Council passed

*Downs, lakes and forests*

a motion demanding that substantially more cable be placed underground in the town itself. If this were accepted, they told Sherrard, all demands for an Eastbourne-type public inquiry would be dropped. But Sherrard refused to concede, and reminded the Council that he believed himself to be in possession of residual powers to bring pylons into the very heart of the town. At this point a compromise was arrived at with the Council settling – once again by a majority of one – for a small increase in underground cabling.[58] But was the 'Keswick war', as it was called by another observer, really over?[59] The answer came early in February 1932, when a 'Keswick Residents' Protest', signed by thirty leading members of the community, and co-ordinated by Kenneth Spence, signalled the beginning of a rearguard action which once again brought the issue of the Grid and the Lake District to national prominence. Sherrard was accused of blackmail: an unnamed 'representative' of the Board was said to have induced 'the council to accept his original offer and to forgo their demand for an inquiry under threat of bringing pylons into Keswick itself'.[60] The activists now found themselves engaged on two fronts. Locally, a growing number of urban and rural district councils were opposed to and impatient with the anti-pylon movement, and the national preservationist lobby was convinced that Spence and his group were destroying the unity which had been built up since the mid-1920s. By spring 1932, there would be a new national initiative and the CPRE would again commit itself to a campaign to keep all overhead cables out of the valley. *The Times*, also, played a part, as it had in relation to the Downland and Lake District schemes three years earlier – there is evidence that the leader writer, H. H. Child, worked on and co-ordinated anti-pylon news and features.[61]

H. G. Griffin at the CPRE believed that his organisation should only act if the National Trust failed to carry out the responsibilities inherent in the Duncan–Bailey accord of 1929. But if relations between the Trust and Keswick could be shown to have broken down – as, by this stage, they undoubtedly had – Griffin would undertake direct negotiations with the Board. Such talks, Griffin added, must allow 'both sides' to be 'fairly heard'.[62] Spence found this difficult to accept. CEB officials, he believed, were openly hostile towards every aspect of the environmentalists' cause. Sherrard could 'talk round nearly everybody': 'he is silver-tongued, glib and very clever, and I fear not at all scrupulous'.[63] But this did not prevent Griffin from inviting Sherrard to a full meeting of the CPRE's executive, while simultaneously warning Spence of the dangers of 'going it alone' and of underestimating the residual power which still lay with local planning authorities committed to the protection of the Lake District.

Here, as on most other issues, Griffin and Spence were in disagreement,[64] and by early March Spence was insisting that 'if we don't get what we want from the Minister then it is up to the CPRE to show that this matter of the Pylons should not be left to the tender mercies of bodies who not [sic] elected to deal with the matter in any shape or form'.[65] Griffin, for his part, accused the Lakeland protesters of having spurned the assistance which the CPRE might have given in 1929 and now, in 1932, of expecting full and unquestioning support.[66]

By May 1932, pressurised by individual members of the CPRE, and by the campaign stage-managed by Child at *The Times*, Griffin sought an exploratory meeting with the Minister of Transport.[67] 'From the point of view of support', he believed it to be 'the weakest case we have ever been asked to tackle . . . and yet we are going to spend more money on it than any other'.[68] But the CPRE in London and the Anti-Pylon Committee in the Lake District were now presenting a united front. Both were now pressing for an inquiry on two counts – that significant numbers of landowners in the Keswick area wished to think again about the wayleave agreements which they had been persuaded to sign; and that the quality of the north-western section of the Grid was suffering as a result of the large amounts of money spent on plans and counter-plans for the inclusion or exclusion of Keswick. 'There seems to be a general feeling getting around', Griffin wrote in October, 'that the whole grid scheme is being carried out in a very extravagant manner and that it will prove to be very uneconomical.'[69] Then, quite suddenly, at the very end of the year, Griffin heard in confidence from *The Times*, that the CEB had decided to exclude Keswick from the National Grid. The formal reason given for the reversal of policy was that demand for electricity in the area was too meagre to justify engineering costs.[70] But a rather different gloss was provided early in 1933 by A. H. Dykes, engineering consultant to the CPRE. 'I have myself', he wrote to Griffin, 'had more than one conversation with Sir Andrew Duncan and Sir Archibald Page and I have good grounds for thinking that they have strong doubts whether it would be to their advantage to proceed with the proposed line through Keswick.'[71] 'Advantage' was an ambiguous word. But it seems likely that the CEB had decided that it could no longer afford to be repeatedly and unnecessarily humiliated by a small town in the North-West.

If the people of Keswick demonstrated a much greater degree of ambivalence and confusion than was ever openly acknowledged by the preservationists, the CEB, vacillating between bombast and inaction, finally compromised both its bargaining power and its credibility. The retreat from Morrisonian *diktat* and its replacement by a policy of administrative defusion and delay was initially well conceived. But this was a strategy which demanded consistency and flexibility. The decision simply to wait for public opinion to turn against the anti-pylon movement was a sensible one. But the frenetic negotiation of wayleave agreements with farmers owning land just outside the town boundaries proved counter-productive. So did Sherrard's over-zealous commitment to electrical progress and his antagonism towards 'time-wasting' preservationists. Confirming prejudice about the arrogance of professional planners and the desire of central government to bypass local democratic processes, Sherrard unwittingly strengthened the resolve of the Anti-Pylon Committee and encouraged the Council to harden its position. The radical environmentalists, expertly organised by Kenneth Spence, were able to convince both the people of Keswick and the national preservationist movement, that they were speaking on behalf of the entire community. The Anti-Pylon Committee could now draw on the support of a wider range of long-established voluntary organisations: the rhetoric of 1929 had been belatedly transformed into the hard currency of collaborative political activism. But, without an insistent rhetorical appeal to history and patriotism, local protest could not have been successfully transmuted into national concern: the Lake District had been presented as the *fons et origo* of a 'sacred' and Romantic tradition of English environmentalism. This stategy, which had not been available to those who opposed pylon construction in the South Downs, was an essential precondition for victory in Keswick.

In the New Forest, conflict over national electrification followed a different pattern. It had at first seemed likely that the CEB would come to a rapid agreement with the three bodies holding a major interest in the region – the Forestry Commission which had administered the area on behalf of the Crown since 1919; the Court of Verderers, representing the rights of landowners and tenants; and the New Forest Association, which was strongly committed to the preservation of an unspoilt environment. The South-Western scheme, including a line between Bournemouth and Southampton, had been uncontroversially adopted in June 1930.[72] Negotiations between the CEB, the Ministry of Transport and the Forestry Commission proceeded well, but when the CEB announced that it intended to erect pylons which would directly jeopardise the integrity of a small but unspoilt area of woodland, both the Commission and the Court of Verderers declared their opposition.[73] The New Forest Association, guided by its president, the former Solicitor-General and

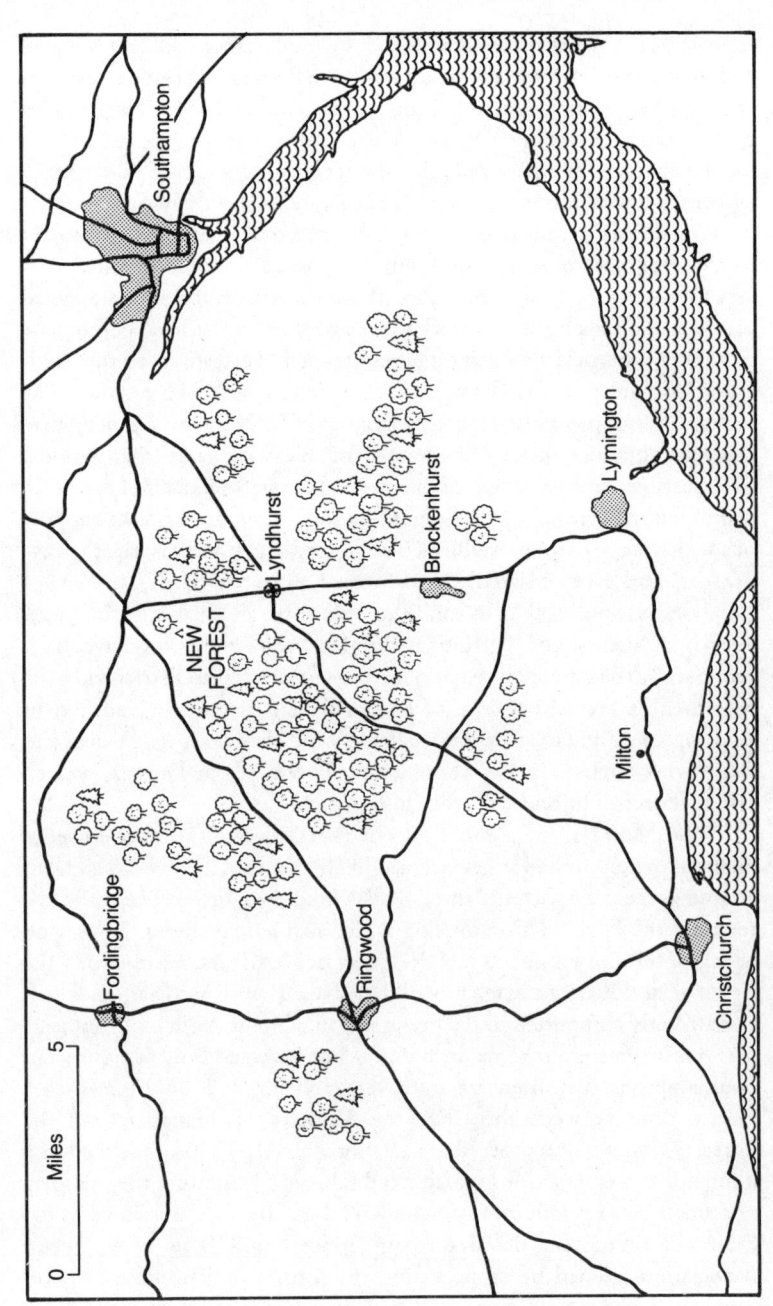

**Figure 8** Foci of electrical debate and literature in the New Forest region, 1930–34

authority on land law, Sir Leslie Scott, followed suit.[74] Having toyed with the idea of taking its cables round the edge of the forest, via an alternative route in the Avon Valley, the CEB now claimed, in the spring of 1932, that non-forest lines would do more damage to the landscape than the original scheme. The New Forest Association immediately expressed its opposition and argued, in late March, that 'further organised opposition may yet be necessary'.[75] But it was the Verderers who first took decisive action, with the Official Verderer, Lord Forster, immediately seeking the advice and support of Stanley Baldwin. 'Nowadays', Forster wrote, 'there is no member of the Government whose duty it is to protect the Forest and so I am constrained to write to you to do what you can to help us.'[76] We do not know how Baldwin reacted to this request, but it is possible that his intermittent commitment to rural preservationism may have played a role in subsequent Cabinet discussions. 'Anti-electric' activism in the villages and small towns of the New Forest now increased. 'From information, letters and reports of Meetings', wrote Cecil Sutton, secretary to the New Forest Association, in early July, 'there is no doubt that feeling throughout the New Forest is very strongly against the proposals of the Central Electricity Board'.[77]

Yet it was only in September that opponents of the scheme called a meeting to publicise their cause. On the eve of this gathering at Brockenhurst, Montague Chandler, clerk to the Court of Verderers, insisted that the proposed departmental inquiry was little more than a cynical 'fishing expedition on the question of the expense of... being allowed across the open lands of the New Forest' as against 'having to go a longer way round'.[78] As for the location of this 'fishing expedition', Chandler wanted to know why 'Winchester had been selected instead of some place within or near the Forest'.[79] Leslie Scott supported this position,[80] and at the protest meeting itself Major J. D. Mills, recently elected Conservative member for the New Forest, claimed that the 'the right place has not been chosen for the inquiry. Winchester! Twenty or thirty miles from the scene of the intended crime! There are numerous places nearby, where it might have been held. If the Ministry really wish to ascertain the views of people of the forest, what about the old town hall of Fordingbridge? That is within five miles of the spot affected.'[81] Cecil Sutton was even more scathing. The Board had little interest in local opinion and, so far as civil servants were concerned, London was the only proper place for the inquiry.[82]

At the hearing itself, Sutton made his points crisply and convincingly. He claimed to speak for 'a membership of no fewer than 1100, consisting of every class of person, from the smallest Commoner to the largest, from the humblest cottager to the mansion-dweller. [The Association] also included lovers of the New Forest in all parts of the country, and even as far as America.'[83] The NFA, Sutton went on, was as much concerned with the manner in which the public inquiry had been organised, as with the potential desecration of one of the most beautiful pieces of natural landscape in Britain. Sutton also claimed that the Ministry of Transport had deliberately kept the hearing away from the very heart of the forest, where a majority were deeply opposed to the intrusion of the Grid.[84] O. A. Sherrard explained why the Board had decided that it must bring pylons inside the forest boundaries. 'It was the business of the Minister', he said, 'to look at these matters from the widest possible point of view, and from the point of view of the largest number and not from the point of view of a small portion.'[85] In Hampshire, as in Sussex and the Lake District, Sherrard displayed a shrewd ability to depict environmental pressure groups as illicit protectors of the interests of a privileged élite which was determined to deprive the urban and rural working classes of cheaper heat and light. In this instance his case was strengthened by the fact that some of Hampshire's richest and most socially elevated residents were now lending their names to the anti-CEB cause. Petitions had been signed by Lady Montagu, the Duchess of Westminster, Lady Curzon Howe and Sir Thomas Troubridge, 'famous for his skill in choosing names for race horses'.[86] The signatories of a letter to *The Times* wrote that 'there are some gifts of Nature that, once surrendered, can never be regained. The New Forest is one of them. It must be preserved inviolate.'[87]

At a time when unemployment occupied more column inches than any other topic in the regional press, such sentiments could only appear deeply reactionary. All this generated a pro-electric, anti-environmentalist backlash. 'So far', a correspondent to *The Times* complained, 'the point of view of the Forest has been totally ignored.'[88] 'What appears undesirable in this controversy', another commentator wrote, 'is that residents in the Forest or any other district affected should claim on behalf of themselves and the public that their own view of the matter should prevail. Such things should surely be decided impartially, and not by pressure of residents.'[89] The electrical press was more direct. 'We think the New Forest Association doth protest too much ... The word "amenities" is getting on our nerves; and to regard a mile or two of transmission line over a corner of the Forest as a danger to the beauty of that expanse is straining the argument to breaking-point.'[90] By early November the 'quite friendly' Forestry Commissioners were reported to be about to begin negotiations with the 'wicked uncle' of the piece, the Ministry of Transport.[91] Simultaneously, Leslie Scott was addressing an all-party meeting of MPs at Westminster and a unanimously pro-environmentalist motion was passed to MacDonald and Baldwin.[92] When the deadlock had first come to the attention of the Cabinet at the end of October, the Minister of Transport, P. J. Pybus, who was not a Cabinet member, had been unable to offer a decisive opinion on the rival claims of the CEB and the united front of organisations determined to protect the forest. He had, however, put considerable store on a report by Sir Reginald Blomfield. Like many another pro-electric enthusiast, Blomfield was convinced that 'the towers would give an added interest to the landscape, and make their own appeal to the imagination'. As for 'non-forest' alternatives, they would involve 'the public [in losing] some beautiful characteristic English landscape which is at present quite unspoilt'.[93]

But the Cabinet, probably swayed by Baldwin and MacDonald, was unwilling to opt for the CEB route. It favoured two interim measures: Pybus should make a personal visit to the forest, accompanied by other senior ministers; and an opinion should be sought from the Law Officers as to the legal claims of the CEB, the Forestry Commission and the Court of Verderers.[94] Pybus asked MacDonald to go down to Hampshire with him, but the Prime Minister pleaded overwork and the visit was eventually undertaken in the company of Sir Roy Robinson, the chairman of the Forestry Commission.[95] The government now hoped to be able to reach a compromise via the good offices of the Commission, and thus obviate potentially lengthy legal confrontations with every body or individual claiming an interest in the Crown Land area. Sir Roy Robinson had retreated from his initial anti-CEB position and, casting himself in the role of mediator, gave an assurance that the Commission would back whichever policy was finally agreed upon in Cabinet.[96] In London, meanwhile, the lawyers were grappling with the possibility of CEB workmen being arrested and arraigned before the Court of Verderers. 'Whether or not the Ministry of Transport', wrote W. A. Brown, the Treasury Solicitor, 'or the Electricity Commissioners are in a position to make an order which, with the passive acquiescence of the Forestry Commissioners, could override the commoners' rights (subject to compensation) is a question of very considerable difficulty.'[97] A week later, Brown's researches had convinced him that 'passive acquiescence' of this type would imply involvement in 'unlawful enclosure, purpresture, encroachment or trespass'.[98] The Court of Verderers might seem to be an anachronistic and self-inflated body, but it possessed inalienable rights;

and the threat that these might be deployed, *in extremis*, against the CEB was an exceptionally powerful weapon in the environmentalist armoury. In the final analysis, therefore, it was unwillingness to see the Board or the Ministry of Transport involved in a long-drawn-out and potentially humiliating conflict wth the Court of Verderers which determined the Law Officers' advice that the government must withdraw.

The anti-pylon campaign in Hampshire had been less intense than in Sussex or the Lake District. This low level of militancy was partly determined by the fact that the New Forest was less radically threatened than the other two regions: the 'corner' which the CEB wanted to incorporate into the Grid could not be compared in aesthetic terms with the noble sweep of the Seven Sisters or the exquisite symmetry of the valley opening out into Keswick. That the CEB was now more willing to compromise had been demonstrated by its readiness to hold a public inquiry at Winchester. Leslie Scott, Cecil Sutton and Montague Chandler may have been convinced that this hearing, like the Eastbourne inquiry before it, was no more than a rubber-stamping of decisions already confirmed months earlier in London. There was much truth in this version of things. Yet growing doubts about the effectiveness of 'administrative defusion' in the Lake District had led to enhanced, though still faltering, commitment by the Board to 'diplomacy' and 'public relations'. (O. A. Sherrard's outbursts were the exception which proved the rule.) What ultimately persuaded the CEB to abandon its more uncompromising plans in Hampshire was the bewilderingly complex legal conflict in which it found itself becoming ever more deeply embroiled. It was forced to concede that the law is a double-edged instrument that can be turned against even the most prestigious public body which itself claims to work from scrupulously legalistic principles.

### Notes

1 *The Times*, 10 October 1929.
2 Leslie Hannah, *Electricity before Nationalisation*, 115–6.
3 *Electricity Commission: Electricity Supply Act: Memorandum on the Provisions of the Act, Prepared by the Electricity Commissioners* (HMSO, 1926), 13–14. See also, Herbert Morrison, 'The Elected Authority – Spur or Brake?' in Royal Institute of Public Administration, *Vitality in Administration* (1957), 14.
4 Public Record Office (PRO) POWE 12/258. 'Report by the Ministry of Transport Inspector, F. Gordon Tucker. September 28, 1929'.
5 *Daily Mail*, 13 September 1929.
6 *The Times*, 13 September 1929.
7 Ibid.

8 *The Times*, 19 September 1929.
9 *Daily Herald*, 27 September 1929 and *Brighton and Hove Herald*, 5 October 1929.
10 *The Times*, 3 October 1929.
11 *Sussex Express*, 4 October 1929.
12 *Sussex Daily News*, 12 October 1929.
13 *Sussex Express*, 18 October 1929.
14 *Daily Herald*, 21 October 1929.
15 *The Times*, 1 November 1929.
16 Ibid.
17 *ET*, 17 October 1929, 593–4.
18 *E*, 4 October 1929, 391. For the alleged involvement of the BBC see *ET*, 16 May 1929, 757. An excellent summary of how the industry believed that it was being mispresented by the media is contained in 'The Central Board, The Press and the Public', *EII*, 9 October 1929, 1658–9.
19 Ramsay MacDonald Papers. PRO 30/69/564. 'Deputations El–Ez'. 14 October 1929.
20 Very nearly every letter received at Downing Street contained this formula.
21 PRO 30/69/564. C. P. Duff to C. Stedman. 29 October 1929.
22 Morrison to MacDonald. 30 October 1929. The highly unconvincing 'composite pictures' – with an artist's impression of the cables superimposed on an excellent landscape photograph – appeared in *The Times* on 28 September.
23 He had, for example, raised the issue of rural conservation during the Halifax by-election. See *The Times*, 16 July 1928.
24 This account of the differing attitudes of MacDonald and Morrison is based on POWE 12/258. 'South Eastern England Electricity Scheme (1927): Deputation to the Prime Minister, November 25.'
25 *Sussex Express*, 22 November 1929.
26 Council for the Preservation of Rural England Archive (University of Reading) CPRE 109/5/1. H. G. Griffin to K. Spence. 23 November 1929.
27 CPRE 237, 'To The Prime Minister', 3. See, also, 'Memorandum to the Prime Minister, and Appendices'. Council for the Preservation of Rural England. *Annual Report*, 1928, 54–8.
28 Herbert Morrison, 'Draft Autobiography', 55. Morrison Papers, Nuffield College, Oxford.
29 *ET*, 17 April 1930, 789. The extent to which major figures in the CPRE renounced confrontation for co-operation in the 1930s is well described by F. R. Sandbach, 'The Early Campaign for a National Park in the Lake District', *Transactions of the Institute of British Geographers*, 3, 1978, 498–512.
30 *ER*, 19 October 1929, 631.
31 *London Evening News*, 16 January 1930.
32 *Architect's Journal*, 2 November 1933. There is an intriguing foreshadowing here of the approach which would later be developed by W. G. Hoskins in *The Making of the English Landscape* (1955).
33 Cited in *The Times*, 11 November 1929.
34 *East Sussex News*, 29 November 1929.
35 One particularly revealing event in the Brighton area was a debate on 22

November on the pylon issue in Ditchling Village Hall between David Edwards, 'an eminent electricity expert', and Lieutenant-Colonel I. H. Powell-Edwards, speaking on behalf of East Sussex County Council. PRO 30/69/564.

36   *East Sussex News*, 6 December 1929. See, also, the comments of A. S. Sutherland-Harris in the same paper on 15 November 1929.

37   On general ideological parameters relating to this topic see Jurgen Habermas, *Legitimation Crisis* (1976, translated by Thomas McCarthy). On the history and development of the public inquiry system, the standard work is R. E. Wraith and G. B. Lamb, *Public Inquiries as an Instrument of Government* (1971). But note the more critical attitude adopted by J. Rodger in 'Inauthentic Politics and the Public Inquiry System', *Scottish Journal of Sociology*, iii, 1978, 103–27 and by Brian Wynne, *Rationality and Ritual: the Windscale Inquiry and Nuclear Decisions in Britain* (British Society for the History of Science Monographs, 3) (1982). On 'centre' and 'periphery' in the context of environmental disputes, see Mary Douglas and Aaron Wildavsky, *Risk and Culture: an Essay on the Selection of Technological and Environmental Dangers* (Berkeley, 1983). Douglas's earlier work – notably *Purity and Danger: an Analysis of Concepts of Pollution and Taboo* (1966) and *Implicit Meanings: Essays in Anthropology* (1975) – has provided historians with a valuable framework for the 'placing' and evaluation of environmental concern.

38   Cumbria Record Office. Electricity Scheme Papers, 1928–9. WDX/422/2/9.

39   CRO WDX/422/2/9. Society for Saving the Natural Beauty of the Lake District. Draft Resolution. 19 July 1928. Lady Harrowby argued that this was a solution which would provide much-needed work for unemployed miners. *The Times*, 19 January 1929.

40   *The Times*, 21 January 1929.
41   Ibid, 8 June 1929.
42   *Cumberland and Westmorland Herald*, 29 June 1929.
43   *Country Life*, 31 August 1929.
44   CPRE 109/16. Kenneth Spence to H. G. Griffin. 17 July 1929.
45   *Cumberland and Westmorland Herald*, 13 July 1929.
46   *The Times*, 20 July 1929.
47   Ibid, 23 July 1929.
48   *Westmorland and Cumbrian Times* and *Westmorland Gazette*, 8 August 1929.
49   *The Times*, 14 September 1929.
50   *Westmorland Gazette*, 26 October 1929.
51   *West Cumberland Times*, 30 October 1929.
52   *The Times*, 2 November 1929.
53   *Cumberland and Westmorland Herald*, 19 July 1930.
54   *West Cumberland Times*, 16 January 1932.
55   *The Times*, 16 January 1932.
56   *West Cumberland Times*, 16 January 1932.
57   CPRE 109/16. Kenneth Spence to H. G. Griffin. 16 January 1932.
58   *West Cumberland Times*, 23 January 1932.
59   *West Cumberland Times*, 30 January 1932.
60   *Cumberland and Westmorland Herald*, 6 February 1932.
61   CPRE 109/16. H. G. Griffin to Dr Leonard Browne. 11 February 1932.
62   CPRE 109/16. H. G. Griffin to Lord Crawford. 2 February 1932.

63  Kenneth Spence to Griffin. 15 February 1932.
64  See the letters between Spence and Griffin, 17 and 18 February 1932.
65  Spence to Griffin. 8 March 1932.
66  Griffin to Spence. 14 March 1932.
67  Griffin to Spence. 5 May 1932.
68  Griffin to Ralph Morton. 16 June 1932.
69  Griffin to A. H. Dykes. 14 October 1932.
70  W. H. Salmon to Griffin. 14 December 1932.
71  A. H. Dykes to Griffin. 26 January 1933.
72  *The Times*, 23 September 1932.
73  Hampshire Record Office. *Minutes and Correspondence of the Court of Verderers* HRO/7 M75 218X. Montague Chandler to Lord Forster. 23 November 1931. On the Verderers in general, see Anthony Passmore, *Verderers of the New Forest: a History of the Forest 1877–1977* (Old Woking, n.d.).
74  PRO CAB 24/234 C.P. 375(32), 'The New Forest and the Electricity Grid. Memorandum by the Minister of Transport', 212–13.
75  HRO 7 M75 267X. Cecil Sutton to Montague Chandler. 17 March and 22 March 1932.
76  Lord Forster to Stanley Baldwin. 5 May 1932.
77  Cecil Sutton to Sir Roy Robinson. 9 July 1932.
78  HRO 7 M75 218X. Montague Chandler to Brig-Gen. E. W. Martin Powell (n.d.)
79  Chandler to Cecil Sutton. 16 September 1932.
80  *The Times*, 20 September 1932.
81  *Bournemouth Daily Echo*, 21 September 1932.
82  *Hampshire Advertiser and Southampton Times*, 24 September 1932.
83  *Bournemouth Daily Echo*, 22 September 1932.
84  *Hampshire Advertiser and Southampton Times*, 24 September 1932.
85  Ibid.
86  *Lymington and Milton Chronicle*, 27 October 1932.
87  *The Times*, 5 November 1932.
88  *The Times*, 9 November 1932.
89  *Lymington and Milton Chronicle*, 10 November 1932. This quotation had been reproduced from *The Times*.
90  *EII*, 28 September 1932, 1511.
91  *Lymington and Milton Chronicle*, 3 November 1932.
92  *The Times*, 4 November 1932.
93  PRO CAB 24/234 C.P. 375(32), 'The New Forest and the Electricity Grid'. Appendix A. 'Summary of and Extracts from Sir Reginald Blomfield's Report', 215.
94  PRO CAB 23 60 (32) 7, 9 November 1932. 'The New Forest and the Electricity Grid'.
95  PRO FI/4. Forestry Commission *Minutes*, 1 December 1933.
96  Ibid, 26 May 1932.
97  PRO MAF, 50/63 A. W. Brown to Sir Roy Robinson. 22 November 1932.
98  A. W. Brown to Sir Cyril Hurcomb. 29 November 1932.

# 7
# Exploiting the Highlands

In Scotland electrical conflict interacted with demands for the indigenous control and exploitation of water power. As early as 1896 the British Aluminium Company had established a hydroelectric plant at Foyers near Loch Ness, and in 1909 had expanded its activities southwards to Kinlochleven, in the vicinity of Fort William. Production soared and, on the eve of the First World War, the company controlled a large proportion of total world output. But in terms of labour conditions and its attitude towards the environment, BAC gained a lamentable reputation. Irish navvies were brought in to work on the construction phase of Kinlochleven and herded into crowded and unhealthy camps; a company housing scheme was built in such a way as to deprive employees of sunlight; and basic environmental desiderata were flouted when the enterprise spread outwards into untouched countryside. In 1921 the company gained additional parliamentary powers to build a hydro-electrically powered factory at Fort William and this was completed in three stages in 1929, 1933 and 1943. Defenders of the scheme insisted that it brought much-needed employment to an area ravaged by economic depression and migration and that the visual excitement of the works high above Fort William increased rather than reduced the flow of visitors. But environmentalists and Scottish nationalists contended that too great an aesthetic price had been paid for no more than a minimal increase in opportunities for local men, and that, in future, every hydroelectric scheme should be subjected to full parliamentary scrutiny under public bill legislation and administered and monitored by a development board situated in Scotland and answerable to Scottish public opinion.[1]

A comparable approach had gained official support in the aftermath of World War I, when the Snell Committee recommended that 'the whole of this potential water power which . . . can be developed for the use and convenience of men at a commercially sound cost is now running to waste. The value of these powers is brought into greater prominence by the continuous increase in the cost of coal.'² In terms of control the Committee advised that 'the water power should either be developed by the State, or leased to public or commercial undertakings for a sufficient number of years to enable the leasees to redeem the large capital expenditure by means of a small annual sinking fund which will not add unduly to the cost of the available power'.³ But economic recession and ideological antipathy worked against any form of state intervention during the 1920s. The Snell report was shelved; successive governments shied away from the difficulties of formulating a balanced, state-backed energy policy; and hydroelectric schemes in rural Scotland continued to be introduced either via English private bill legislation or the labyrinthine Scottish bill procedure. (If the latter route were chosen, evidence was frequently heard behind closed doors in Edinburgh, and, when recommendations were brought before Parliament, the Lords could play as large a role in shaping legislation as the Commons.⁴) By the end of the 1920s a hard core of Scottish members had become convinced that London had little interest in promoting hydroelectric schemes which were relevant to the needs of farmers and crofters, or in delegating responsibility either to the Scottish Office or to the Central Electricity Board. 'Scottish power' was in their eyes being exploited in an *ad hoc* manner by wealthy and alien companies which expended large sums of money in legal fees to persuade select committees that their intentions were honourable and progressive, but then did little to tailor their plans to the small-scale electrical needs of Highlanders or to protect a uniquely resplendent landscape. This was the context in which Scottish members, supported by powerful sporting and preservationist interests in England, voted down seven attempts to harness water power in the west Highlands between 1928 and 1941.⁵

David Kirkwood gave classic expression to this anti-English animus in 1929 when he argued that 'it is our duty to see to it that this new power of electricity is kept in the hands of the community, and not allowed to get into the hands of exploiters. Those individuals who are running this show do not care if they run Scotland dry as long as they make money out of it.'⁶ George Hardie, younger brother of Keir, expressed the same set of ideas when he stated that 'we are fighting against the right of any handful of individuals in any form of ramp to have handed down to them a natural source of energy called water, in order that they may sell back power to

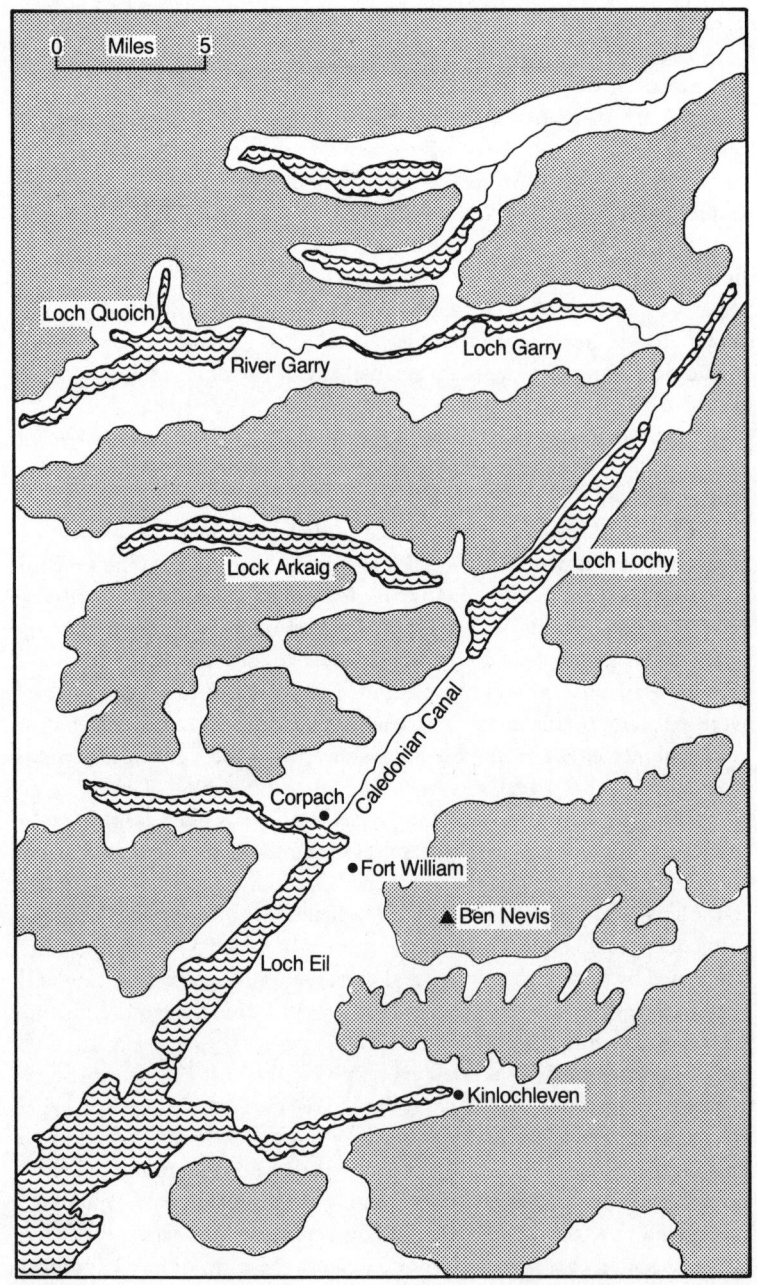

**Figure 9** The area of Inverness-shire most directly affected by the Caledonian schemes

this nation at a profit'.⁷ Full confrontation, involving the British Oxygen Company, Scottish nationalists, environmentalists from both sides of the border, and the National Government, was precipitated by the Caledonian Power Scheme which came before Parliament under private bill procedure on no fewer than three occasions between 1936 and 1938. The object of each of the bills was to obtain permission to construct a hydroelectric power station and a carbide factory at Corpach near Fort William in a wholly unspoilt region of Inverness-shire. (Calcium carbide is a chemical which played an important role during the First and Second World Wars for oxy-acetylene welding and munitions production.) The proposal involved five dams, between 30 and 60 feet high, and four additional power stations, connected to one another, and linked to Corpach by pylons and overhead lines, stretching for more than 50 miles. Thirty miles of cable would run directly down the glen in which the Caledonian Canal was situated; the natural level of several lochs would be radically altered; and the flow of numerous rivers, large and small – notably the Garry and the Moriston – would also be affected.⁸ As soon as the project was announced, in November 1935, the Association for the Preservation of Rural Scotland sent a circular to Scottish MPs, pointing out that, in addition to the environmental threat posed by the new scheme, existing hydroelectric development in Scotland had been much less successful than financially interested parties had claimed. Coal, rather than water, cried out to be exploited during a depression which was decimating Scotland's long-established industrial base and the performance and potential of Scottish hydroelectricity needed to be fully reassessed before the government allowed any further major companies to begin construction in the glens.⁹

In the event, lobbying by the APRS, together with Scottish and English suspicion of the financial and rating terms under which BOC would be allowed to start operations, ensured that the first Caledonian Power Bill was defeated by 199 to 63. Archibald Ramsay, the member for Midlothian and Peebles, who would later be imprisoned for four years during the Second World War, for making anti-semitic statements, gave voice to a strong body of opinion. He objected to the rating remission which would be granted to the company and insisted that 'schemes of this kind, instead of going to employ Irish Labour in navvying in odd parts of Scotland, destroying the beauties of the countryside, annoying every section of the community and destroying amenities [should] go to parts of Great Britain where they are needed'.¹⁰ The people of Inverness, who held that the diversion of the River Ness would irreparably damage their sewage disposal system, were said to be 'very anxious not to be compelled to engage expensive counsel and technical experts to defend their rights ... in view of the fact that they have already had to do the same thing twice within the last six or seven years'.[11] This was a point which was reiterated by Sir Murdoch Macdonald, the town's MP, who contended that the scheme was likely to deprive Inverness of 'sixteen per cent of the water now flowing down through that beautiful river ... one of the most beautiful rivers in Scotland, if not the most beautiful, is going to be wholly taken from them'.[12] John Davidson, the radical member for Glasgow, Maryhill, pointed to the potentially unconstitutional character of a bill which tied rating concessions to a monopoly position in terms of electricity supply to rural areas. 'Here we have the House', he complained, 'being asked by a private concern to allow it not only its own particular rating, but to allow it to control the local authorities in such a way that it will be able to fix the price of electricity in that particular area."[13]

As for members who supported the bill, one stated that its opponents 'had cried out for years for industry to come to the aid of the Highlands but that, now help was at hand, they had dragged up every kind of trivial reason to oppose such an initiative'.[14] In terms of environmental and sporting interests an English Labour member was confident that 'these objections in regard to amenities come from the people of leisure who do not like the amenities of their particular district disturbed. We ought to regard this country as a treasure house of the people rather than a pleasure ground of the leisured class."[15] Thomas Cooper, the Lord Advocate, did little more than attempt to persuade the House to refer the bill to a select committee in order to resolve the rating problem. He neither overplayed the strategic importance of carbide production nor insisted that the country must be adequately stocked with domestically manufactured rather than imported supplies during time of war. But he did remind members that the issue should be considered from a 'United Kingdom standpoint'. As for the argument that coal rather than water should be developed during a period of traumatically high unemployment, Cooper merely noted that 'it is not a question of whether the industry should be located in Inverness-shire or Lanarkshire, but whether we should have the industry in Inverness-shire or not have the industry at all'.[16] Within weeks of the defeat of the first bill, BOC let it be known that it would bring a revised version before the House at the earliest opportunity. The company was confident that it could resolve the rating problem through direct negotiation with interested parties. It was also convinced that it would be able to mollify those who argued that the carbide factory should be sited in a Special Area, by tying the Corpach project to another, steam-powered factory in South Wales.[17] By this juncture, also – the early months of 1937 – the government had hesitantly and unwillingly come to the conclusion that the strategic implications of the Caledonian Scheme dictated that it must decide what attitude should be taken to the re-presented bill. As Minister for the Co-Ordination of Defence, Sir Thomas Inskip minuted that he had 'no doubt that the Bill will be defeated on Second Reading unless a very emphatic convincing statement is made by a Government spokesman in support of the Bill'.[18] But, if the government were to become more directly involved in the controversy, it must bear in mind that BOC would seek protection against imports of carbide from Norway and a subsidy for the South Wales factory which went far beyond the usual levels of state aid. Cordial Anglo-Norwegian trading relations, with Norway exporting large quantities of agricultural produce for British Coal, could well be disrupted.[19] Walter Elliot, the recently appointed Secretary of State for Scotland, argued in favour of unequivocal commitment to the Corpach scheme.[20] Ernest Brown, the Minister for Labour, took a different view. Special Area status had not yet attracted nearly enough new industry to South Wales, and since the absolute number of unemployed in the Highlands was minute when compared with comparable figures for the heavy industrial regions, the government should back the steam-powered option and beat down BOC's financial demands when the Scottish scheme had been dropped.[21]

When the issue came before the Cabinet in February 1937, none of these arguments prevailed. There was a consensus that a domestically produced supply of calcium carbide was not strategically vital: Canada manufactured a massive 150,000 tons a year and, if, during a hypothetical conflict, the North Atlantic sea-lanes could no longer be controlled by the Royal Navy, then the war would already have been as good as lost.[22] Sir Thomas Inskip reflected this agnostic and, as we shall see, unrealistic view, when he wrote at the end of February that 'if the seas are not kept open, the loss of calcium carbide will not rank very high in the evils which would follow'. He added that, in terms of safety from air attack, neither the Highlands nor South Wales were more or less vulnerable; and that in terms of cost, the Electricity Commissioners had confirmed that coal and hydroelectricity were broadly comparable.[23] As for the 'Norway connection', Walter Runciman, at the the Board of Trade, reminded the Cabinet that negotiations were in train to improve relations *vis-à-vis* agricultural products and that Norway, which supplied more than 90 per cent of its total annual production of carbide to Britain, would not look kindly on a subsidy for the proposed BOC factory in South Wales. It might, in fact, cut back on its imports of British coal – and that would be disastrous.[24] Nor had any single view gained dominance in Cabinet by early March. A meeting on 3 March revealed a continuing unwillingness to think about the unthinkable – the possible closure of the Atlantic sea lanes during time of war. There was general agreement that the bill could not be carried unless it were fully endorsed by the Cabinet and that, if English members played a decisive role in outvoting it, the government might find itself politically embarrassed. But any form of unambiguous endorsement was considered to be unwise and Walter Elliot was briefed to keep every option open. This was reflected in the advice that 'members of the Government should not vote against the Bill but . . . each member should be free to decide whether he voted at all'.[25] This vacillation would in due course sow confusion in the ranks of the government, undermine the credibility of both the Corpach and the South Wales options, and inhibit the development of a viable carbide policy when war did break out.

By March 1937 both the APRS and the CPRE were capitalising on BOC's decision to link the Corpach development to a second factory in a Special Area. Not only would the Highland project threaten the livelihood of communities which were heavily dependent on tourism, ghillying and odd-jobbing for the privileged few who either lived in or visited Inverness-shire to shoot and fish; the same end-product could be obtained at comparable cost in a traditionally industrial area.[26] Tension was also mounting in the Highlands themselves with mass meetings being held in Glengarry, Fort Augustus, Glenmoriston and Inverness.[27] The Reverend James Hill, Minister of Glengarry, took the exceptional step of addressing protesters from the pulpit. 'This scheme is merely changing the type of landlord', he told them, 'making the directorate of a company the new landlord.'[28] In a letter to *The Times*, Lord Portland quoted a letter from the great-great-great-granddaughter of the last chief of the Macdonells. 'There are hundreds – probably thousands – of Highlanders in Canada and other parts of the world who think of Glengarry . . . and Glen Moriston as home.'[29] But there was also good news for BOC: the Parliamentary Bills Committee of Inverness County Council had been won over. 'It is difficult . . . to see why this industry which, if established in the Highlands will cost the Treasury nothing', the Committee's chairman noted, 'and be of considerable benefit to the Highland rate-payers, should be bribed away to another Special Area, where it will cost the taxpayers over 100,000 a year.'[30]

When the bill received a second reading for a second time, its opponents were angered by BOC's refusal to acknowledge the depth of local feeling. Archibald Ramsay contemptuously dismissed the new rate rebate of 33 per cent for a period of five years and denied that the project would stimulate Highland industry. It was more likely in his view that Irish navvies would be brought in for the constructional phase and then stay on as a permanent burden on the rates.[31] John Davidson repeated his socialist critique:

Created in 1886 as a small company, today the issued capital is over £3,000,000. This company is asking for rating reductions more than the House has ever agreed to any ordinary shopkeeper or householder, yet it had paid £1,900,000 in profits during the past 10 years. Since when did it become our objective, as members of a Socialist party, to support the inauguration of an industry on such terms as this, which will inflict harm on the town of Inverness?'[32]

The British Oxygen Company had cannily recruited Robert Boothby to speak on its behalf and in an eloquent speech, which revealed his profound disapproval of the government's rearmament policy, he attacked both the preservationist lobby and those who argued that the adoption of the Highland scheme would not lead to large-scale social and economic regeneration. Boothby reminded the House that the Highlands had been losing population at a rate of 4000 a year since the end of the First World War. As for those who were left behind, 'from the way some people talk and write', he stated, 'one might imagine that the crofter's life was a kind of idyll. It is probably the hardest life in the world, and since the War it has become desperate. Why not give these men a chance to earn a decent living in a permanent way?'[33] In no circumstances, Boothby went on, should members be swayed by 'occasional' Scots – 'these owners of sporting rights in the north of Scotland [who] have suddenly evinced this intense and almost passionate interest in the problem of the distressed areas, particularly South Wales'.[34] Both the crofting community and the magnificent environment which it inhabited could be assured of protection and an acceptable solution would also be found for the rating problem. Two further issues – the long-term economic revival of the Highlands and national defence, during a period of increasing international tension – were crucial. The real question in this respect, Boothby argued, was whether hydroelectric and related chemical industries were to be established as a long-term investment in this country or whether we would continue to be foolish enough to rely on foreign supplies. 'It is not a question of Corpach or South Wales', he warned. 'The decision is between Corpach and Norway.'[35] More than any other member of the

House, Boothby starkly revealed the indecisiveness and lassitude which characterised the government's attitude towards the calcium carbide issue. This was brought into clear perspective when Walter Elliot tried to tread the narrow line between acknowledging the close relationship between the Corpach scheme and national defence, and denying open support for BOC's application. 'We trust', he told the House, that 'this country will continue to have command of the seas and to have access to supplies from foreign countries'. But were the sea lanes to be closed, it was 'one of the most improbable things that could possibly happen, that the Government should take upon themselves to set up a factory to make carbide'. If, though, the bill were to be defeated, the government would not be able to 'divest itself of responsibility ... it would take up the question of how best the country could be provided with a supply of calcium carbide and where best that supply of carbide could be produced'.[36] It was small wonder, in the light of ambiguities such as these, that Boothby demanded impatiently, 'Where does the right hon. gentleman stand?'[37]

The day after the second Caledonian Bill was voted down by 256 to 138, Boothby accused English and Welsh members of having maliciously thrown out a measure which was vital to the future of Scotland.[38] This was an over-simplification. The bill was lost because the Cabinet was unwilling to tell the House that an independent calcium carbide industry was of vital strategic importance. Had Elliot been allowed to make such a statement, the bill would have been approved without further ado. As it was, Boothby's speech played an important role in forcing the Cabinet to set up a secret committee, under Inskip's chairmanship, to review the entire carbide position and to determine 'whether the need of home production ... is sufficiently strong on Service and industrial grounds to justify the imposition of a duty and the consequential amendment of the Norwegian Trade Agreement under which the article is now imported free'.[39] The Carbide Committee heard more than a dozen major witnesses but BOC's expert presentation of its case, together with the experience it had gained through close co-operation with the Norwegians over a long period of time, ensured that it won the day; and, by August, the government appeared to be putting its full authority behind both the Corpach and the South Wales initiatives. (The latter was now to be sited at Port Talbot.)

But if it seemed, in industrial circles, that Inskip had finally committed the government, Baldwin was, as always, ready to indulge in the politics of delay. In a crucial Cabinet meeting in mid-November 1937, he informed his colleagues that he did not agree that the establishment of a calcium carbide industry in this country was essential to defence, though it might be advantageous. Apart from the Scandinavian source, supplies were available in other parts of the world, for example in Canada, and could, if necessary, be brought by sea. Even if, however, one factory was desirable from the point of view of defence, why was it necessary to have two? Very strong objections were taken in Scotland to the factory at Corpach on the ground that it could ruin amenities of some of the most attractive districts of the Highlands. He did not see why, if a factory were established in South Wales, the British Oxygen Company should be allowed to go ahead with a factory in Scotland as well.[40]

Baldwin might well have been speaking on behalf of the Highland preservationist lobby. If, as Inskip's Committee had concluded, a native carbide industry was crucial to national defence, and if the Port Talbot project could start production more rapidly than any hydroelectrically powered equivalent, the plans for Corpach had become redundant.[41] 'Members who oppose the Scottish project', *The Times* pointed out, 'will argue that, now it is admitted that calcium carbide can be produced in South Wales on a commercial basis, the whole scheme should be concentrated there.'[42] But BOC chose to ignore the political logic of such arguments, and, by the spring of 1938, it was pushing ahead on several related fronts: arranging the generation of its own electricity at Port Talbot; tackling the thorny problem of the Scandinavian carbide quota; and attempting, yet again, to win over Highland opinion.[43] But with the third Caledonian Bill due to come before Parliament in early April, the company's confidence was not matched by any unanimity on the part of the government. When an anxious Inskip sought guidance from his superiors, he was informed by Baldwin and Sir John Simon, the Chancellor of the Exchequer, that no firm commitment should yet be given. The British Oxygen Company must assure the Cabinet of its *'bona fide* intentions to proceed immediately with the establishment of a factory in South Wales' but 'any overstatement of the importance from the point of view of defence of the establishment of these carbide factories' should be scrupulously avoided. The government must present itself as being anxious to listen to 'the *desiderata* of those concerned in the amenities of the Highlands'. The Cabinet must, above all, be granted yet further breathing-space to formulate a convincing carbide policy in the eventuality of war.[44]

Meanwhile, the APRS and CPRE continued to hammer away at the contradictions inherent in the Carbide Committee's now much-publicised report. Information on comparative costs, according to Kenneth Spence, pointed firmly to Port Talbot and ensured that there was 'no argument left for those who would . . . ruthlessly destroy for ever the glories of this

Highland scheme'.[45] The same point was made by the leadership of the APRS. They claimed that calcium carbide was the 'magic phrase that is to justify all the evils' involving fifty miles of Highland glen scenery being replaced by 'a sordid collection of reservoirs, dams, pipelines, power stations, and pylons'. Why should any of these despoliations be carried out now that the Port Talbot scheme had been officially admitted to be cheaper and, in terms of reducing unemployment in one of the most notorious of the Special Areas, more labour-intensive than in the Highlands?[46] Preservationist and nationalist arguments were restated with increased intensity. 'In the name of honest speech', wrote a pamphleteer on behalf of the radical Highland Development League,

let those who talk of the Development of the Highlands cease to write of these chemical and metallurgical industries, which come with the large-scale development of water power, as if they opened a door into a Promised Land. Not a Promised Land nor a New Testament therein, but a life of growing misery and of physical and mental degradation wait on those who enter . . . if such things were planned for the Lake District the whole of England would rise in furious protest; yet the Scottish area which is thus threatened is as large as the whole Lake District and immeasurably more beautiful.[47]

'The first step', an APRS hand-out insisted,

may be through the passing of an Act making provision for a special Highland Electricity Board, to have full control of the production and distribution of electric power within the Highlands and to be guided by the principle that this power shall be used primarily to encourage the development of industries and occupations likely to lead to a real and widespread revival of Highland prosperity.[48]

Immediately before the third debate in Parliament *The Times*, that bastion of moderate preservationism, attempted to weigh up the pros and cons of large-scale hydroelectric development in the Highlands. 'The controversy', it noted, 'convulses the district in the catchment area of the waters that flow into the Caledonian Canal. Inverness in opposition, and Fort William in support, are waging an argumentative war; and in Glen Garry a minister has prayed against the Bill from a pulpit. 'There was, the paper concluded, little that could be termed genuine 'vandalism' in the proposal – controversy probably centred less on environmental issues than 'regret over the decay of life in the Highlands . . . interest in the production of an important material of war, and . . . hope that a Special Area will be given employment'.[49] 'Among Scotsmen', another *Times* editorial insisted, the main argument is between those who see in a carbide factory at Corpach a desecration of a sanctuary of natural loveliness, and others who, thinking of men rather than the glens, welcome the prospect of employment and economic betterment which a factory would bring to impoverished people now maintaining a mean existence on small crofts . . . But if the Highlands are saved the Highlander ought not to be sacrificed. What shall be done for the men of the Highlands who cannot flourish on old romance or on an unspotted natural beauty?[10]

In terms, finally, of the repercussions of the Carbide Committee's conclusions on the likely outcome of the vote in Parliament, the paper stated that

the promoters, by offering to erect a factory in South Wales, have admitted that calcium carbide can be produced economically without subsidy within a much shorter period in an area where steam-produced electricity can be obtained, and it is submitted [by the bill's opponents] that it would be more advantageous to erect a factory in a depressed area, where all the materials for the production of electricity and calcium carbide are available on the spot.[11]

On the eve of the debate, Scottish supporters of the scheme made a final attempt to influence the vote in the Commons. 'In another generation', predicted Flora MacLeod of MacLeod,

this wonderful and passionately loved country will be in the condition which apparently in the eyes of the lovers of scenery is so infinitely desirable and romantic. The glens and mountains will be deserted. They will become the playground of the country during two, three or four months and be left to hotel-keepers and deer for the rest of the year.[12]

Inverness County Council expressed its firm support for the scheme; and so, also, did two-thirds of those attending a meeting of residents at Glengarry.[13] There was evidently no connection between living in the Highlands and wanting to see the region indefinitely insulated from every form of development.

During the Commons debate, opponents of the project returned to the now-familiar themes of the public control of water power and the inviolability of the municipal rights of Inverness. Murdoch Macdonald reminded the House that

the Bill asks that a private company, for its private gain, should be allowed to inflict injury on the town of Inverness and its inhabitants . . . The day of the development of Highland water power is bound to come but when it does come let us have it in such a way that the maximum injury will not be done by private people for private gain.[14]

John Davidson referred again to 'this wealthy firm, which has shown such tremendous profits [and] asks for a monopoly over the water power of the Highlands and for power to destroy hundreds of miles of scenery, for the employment of 300 men'.[55] Large numbers of English members opposed the bill on environmental and aesthetic grounds. But a minority of Scottish MPs – probably about a third of the sixty attending each of the Caledonian debates – still rejected the project because they disapproved of an inherently 'English' private bill procedure which disposed of the natural resources of Scotland to non-Scottish consumers. Localism, nationalism and socialism combined against the London establishment, and, to a lesser extent, its vassal, the Scottish Office in Edinburgh, which so consistently turned a blind eye to the traumas of unemployment and social deprivation experienced both by the industrial proletariat and by Highland farmers and crofters. The electricity and 'water rights' issue had become potently symbolic of a deep political schism between England and Scotland.

Like the socialists and nationalists, Robert Boothby, again persuasively supporting the bill, emphasised that the far North-West was suffering an extended period of intense social and economic deprivation. 'The Highlands are not merely a distressed area' he said, 'but a derelict area.'[56] Part-time employment, directed largely towards satisfying the needs of an Anglo-Scottish shooting and fishing élite had, in his view, undermined local initiative, and contributed to the stereotype of the Highlander as an individual who 'would not like good steady employment at a reasonable wage'.[57] Only radical action – carefully planned investment in the economic infrastructure and small-scale, flexible industries fitted to the environment and geography of the region – would reverse long-term, secular decline. Such interventionism, of which the Caledonian Scheme marked no more than a beginning, might well disrupt the pleasures of a patrician minority. 'Far be it from me to lay my hands upon the sacrosant and rights of the sporting fraternity', Boothby half-joked, 'but there are occasions when not many are involved, when we have just to grind our teeth and face up to it.'[58] The price that 'would have to be paid in terms of social disharmony was small when one remembered that the Corpach project would require between 3000 and 5000 men during the constructional phase and 300 to 500 in the long-term'.[59] Sir Thomas Inskip found himself torn between a need to follow the continuingly vague Cabinet line and a desire to reassure the House that BOC was now more ready than ever to invest heavily both in the Highlands and in South Wales. 'By pooling the costs of the two schemes', he announced, 'the company expects to be able to manufacture and to distribute at a reasonable rate of profit . . . That is why the company have put the two parts of this scheme together as parts of a complete entity.'[60] 'We are the only important nation in Europe which does not at the present time manufacture calcium carbide', Inskip continued, 'I suggest it is an advantage that private persons should be prepared to put up the substantial sum of money to establish this industry in our country.'[61] But when he was confronted by Tom Johnston over the strategic importance of domestically produced supplies of carbide, Inskip immediately retreated, like Elliot before him, into impenetrable qualification. 'I do not take the responsibility of saying that I do not believe that it is absolutely essential to Defence that this calcium carbide and these ferro-alloys should be manufactured in this country . . . [But] I can only say that the passage of this Bill . . . would be a great relief to me in my anxious responsibility and would make a most valuable contribution to the solution of our supply problems in time of war.'[62] A majority of English members would have been ready to drop their commitment to preservationism and Scottish sport if they had been simply and firmly told that His Majesty's Government considered the establishment of a calcium carbide factory to be vital to national defence. But what they heard and what finally defeated the third Caledonian Power Bill by 86 votes was Inskip propping up a vacillating Cabinet, while simultaneously pleading with the House to ease his own position by lending support to a private bill.

'Nothing', *The Times* reflected the next day, 'but the decisive intervention of the Government on the side of the Company would have sufficed to persuade the House of Commons to pass a private Bill which threatened to change for the worse so many fine and desirable things.'[63] *The Electrician* was less diplomatic. 'The half-hearted attitude of the Government in the matter seems extraordinary and the proceedings in Parliament reflect a confusion of thought and purpose which is altogether regrettable.'[64] The *coup de grâce* came two months later. Keeping rigidly to the position that Corpach and Port Talbot were integrally linked, BOC informed Inskip that 'after full consideration of the situation and especially of the present commercial possibilities for the production of calcium carbide, they did not feel justified in proceeding with the Port Talbot part of the scheme'.[65]

The spectre of the Caledonian Power Scheme haunted successive governments for the next five, tumultuous years. In 1936 and 1939 spokesmen for Chamberlain's ministry were pressed on the calcium carbide issue but gave very little direct information, while simultaneously deploring the House's failure to endorse the private bills of 1936–8 and reminding members that a disastrous precedent would be established if public money were to be spent on a government-backed factory.⁶⁶ Then in September, 1939, Colonel John Llewellin, Parliamentary Secretary at the Ministry of Supply, told the House that the government was 'well aware of the importance of calcium carbide and that if it becomes necessary we shall certainly set up an industry . . . at the moment we have a large store of that material in this country'.⁶⁷

Following the fall of Norway the issue was raised again, with Malcolm MacMillan unrealistically suggesting to Herbert Morrison, the Minister of Supply, that a hastily constructed hydroelectric plant might ease any shortfall in national carbide requirements. Morrison replied that supply 'had been provided for some time to come'.⁶⁸ Then, in May 1943, H. J. S. Wedderburn, member for Renfrew, spoke of the government's 'tragic mistake' over the Caledonian Scheme and claimed that shortages of carbide had been a serious impediment to war production.⁶⁹ Three months later Sir Andrew Duncan, who was now Minister of Supply, calmed members' fears by revealing that there were in fact two state-backed factories in the country; but, he went on to add, in classic Baldwinian style, that the government would withdraw from all such employment-boosting projects when hostilities ceased.⁷⁰ So far as one can tell, government plans for ensuring that the war effort would not falter for want of adequate reserves of carbide were moderately successful; there seems little doubt that, following the fall of Norway, it was Canada which played a key role as supplier of the chemical. But it was precisely now in 1940 that the 'sea lanes' issue, first raised in 1937, became a reality. Canadian carbide was transported by convoy, and convoys required incessantly vigilant and costly protection. Nothing is known about the ratio between tonnages of specific categories of war material loaded and tonnages delivered, or about loss of life attributable to the import of strategic goods which could just as easily have been manufactured at home. When Robert Boothby wrote, in 1947, of 'vital supplies' of carbide being transported 'at great risk and great cost in lives and money', he was referring to the North Atlantic crossings.⁷¹

In Scotland, meanwhile, war-time consensus and Churchill's willingness to grant a relatively free hand to Tom Johnston, the new and able Secretary of State, greatly reduced the social, political and environmental tensions inherent in the control and exploitation of hydroelectricity. Accepting the recommendations of a committee chaired by the former Lord Advocate, Lord Cooper, Johnston placed his full authority behind the creation of the North Scotland Hydro-Electric Board.⁷² When he introduced the necessary legislation in February 1943, the Secretary of State, like Boothby before him, drew attention to the traumatic demographic decline that had occurred in the Highlands during the last two generations. In 425 selected rural parishes the population had fallen by about 100,000 between 1901 and 1931; in Caithness, Sutherland and Shetland, one-third of the total number of inhabitants had been lost over a period of 60 years.[73] Like Boothby again, Johnston was convinced that hydroelectricity was crucial to any form of genuine economic regeneration. As for the question of control, he told Scottish members that 'so long as the schemes come to Parliament, and so long as there is a Minister responsible for them to Parliament, public interest can be adequately guarded'.[74] Confronted by David Kirkwood with the criticism that the embryonic Board would still be too heavily dominated by and responsible to London, Johnston retorted, 'I am trying to show ... this is the voice of Scotland for once'.[75] This was no time, he implied, for grandiose plans for ideologically pure forms of nationalism or socialism – he was determined to obtain as much as possible while the English were otherwise preoccupied, but there must be no 'excesses', and hydroelectricity should not be deployed either to define or press for any form of 'independence'. 'The Bill will give considerable employment', he assured the once-cantankerous but now much more receptive Scottish benches, 'direct or indirect in coal, iron, steel, cable-making, electrical engineering, cement, house and civil building works and contracting.' Increased labour demands in the Highlands were optimistically estimated to be in the region of 10,000 men 'for a number of years'.[76]

Viewed from the perspective of the 1950s, preservationist and 'control' conflicts over Scottish hydroelectricity, which had reached such high levels of intensity during the 1930s, had lost much of their relevance. There were few complaints about aesthetic vandalism in the glens. During his chairmanship of the Board between 1946 and 1959 Johnston ensured that some at least of the promises that had been made to Highlanders during the war were fulfilled: cheaper electricity was made available to larger numbers of people in remote areas and the new source of energy was used to power flexible and relevant forms of industry. According to Leslie Hannah, 'the long debate between landowners, industrialists and others about who, in effect, should appropriate the economic rent of potential hydroelectric sites, had essentially been resolved in favour of the people living there (to the extent that they were willing to take it in the form of subsidised electricity at home) and this indeed had been Johnston's intention.'[77] But, for the Highlands and for

Scotland as a whole, the Board created as well as solved long-term structural problems. A revisionist historian has recently stated that Johnston perceived the new body 'as an agency for regenerating the Highlands, a national investment, and a triumph for Scots autonomy ... But it was none of these things; without it coal might not have hit the trough of the early 1960s.' Nor, from the mid-1960s, the same author continues, was hydroelectricity able to meet a greatly increased national demand; a new generation of coal, oil and nuclear stations had to be brought on-stream to correct the deficit.[78] By this juncture, also, the Highlands and Islands were confronted by new and subtly different social and economic dilemmas. Johnston and his fellow moderate autonomists had been careful to avoid the claim that hydroelectricity *per se* would or could revolutionise life in the glens. But the new energy source probably achieved much less, in the long term, than that generation of optimistic politicians and administrators hoped or expected: too much damage, in terms of massive economic exploitation and demographic attrition, had already been done. Surveying the 'Highlands and Islands problem' in the early 1990s one is inclined to agree with the pessimistic conclusion of the economist Adam Collier, who wrote, at the height of the Caledonian controversy, that there always seemed to be 'such a mountain of exploitation of Highland resources and such a mouse of social improvement'.[79]

### Notes

1  This account is based on David Turnock, *Patterns of Highland Development* (1970), 159–62; R. A. S. Hennessey, *The Electric Revolution* (1972), 95–101; and Hugh Quigley, 'The Highlands: an Economic Question', *EA*, January 1937, 171–3 and 'The Highlands of Scotland: Proposals for Development', *Agenda: A Quarterly Journal of Reconstruction*, 3, 1944, 77–96. See, also, the bibliography in Charles Loch Mowat, *Britain between the Wars: 1918–1940* (1955), 469–70.

2  *Interim Report of the Water Power Resources Committee*. PP 1919, XXX, 833.

3  Ibid, 834.

4  For a revealing dispute on the alleged secretiveness of the Scottish Bill procedure, in relation to the water power issue, see *Hansard*, 372, 25 June 1941, 1054–60.

5  Hugh Quigley, *Agenda* (1944), 84.

6  *Hansard*, 227, 25 April 1929, 1168–9.

7  Ibid, 1174 and 1180.

8  CPRE 102/8. H. G. Griffin to Gordon Winter. 20 March 1936.

9   CPRE 102/8. Hon H. W. T. Scott to Sir Godfrey Collins. 1 January 1936. See, also, the APRS handout, 'Caledonian Power Bill' (n.d., but probably, March 1936).
10  *Hansard*, 310, 18 March 1936, 520–1.
11  Ibid, 523.
12  Ibid, 526.
13  Ibid, 561.
14  Ibid, 555. Joseph Maclay, member for Paisley.
15  Ibid, 565. Francis Broad, member for Edmonton.
16  Ibid, 548.
17  CPRE 102/8. Edward McGregor to H. G. Griffin. 24 March 1936.
18  CAB 24/268/55. 'Caledonian Power Order, 1937. Production of Calcium Carbide. Memorandum by the Minister for Co-Ordination of Defence', 3. 9 February 1937.
19  Ibid, 4.
20  CAB 24/268/60. 'Caledonian Power Order, 1937. Production of Calcium Carbide. Memorandum by the Secretary of State for Scotland'. 11 February 1937.
21  CAB 24/268/62. 'Caledonian Power Order. Production of Calcium Carbide. Memorandum by the Minister of Labour', 1–4. 15 February 1937.
22  CAB 23/8(37)5. 17 February 1937.
23  CAB 24/268/72. 'Caledonian Power Order. Production of Calcium Carbide. Memorandum by the Minister for Co-Ordination of Defence', 3. 26 February 1937.
24  CAB 24/268/79. 'Calcium Carbide. Memorandum by the President for the Board of Trade'. 1 March 1937.
25  CAB 23 10(37)3. 3 March 1937. During the First World War calcium carbide had been imported, first from Norway, and later from Canada. But convoy costs had become prohibitively high and the government established a factory at Spondon in Derbyshire. This plant later fell into dereliction in the face of foreign competition. See the comments of Brigadier-General Sir W. Alexander, who had been Controller of Aircraft Supply and Production at the Ministry of Munitions, in *Hansard*, 310, 18 March 1937, 528–30.
26  CPRE 102/8. 'Caledonian Power Bill' (Circular to members of both Houses of Parliament). March 1937. Comparative unit costs had been estimated as 0.222d for hydroelectricity and 0.152d for steam-produced supplies. See 'Caledonian Power Bill' (APRS). March 1937.
27  *The Times*, 3 March 1937.
28  Ibid, 4 March 1937.
29  Ibid, 9 March 1937.
30  Ibid, 9 March 1937.
31  *Hansard*, 310, 18 March 1937, 1237–43.
32  Ibid, 1269.
33  Ibid, 1256–7.
34  Ibid, 1258.
35  Ibid, 1260.
36  Ibid, 1278.
37  Ibid, 1279.

38  *The Times*, 12 March 1937.
39  CAB 24/269/100. 'Calcium Carbide. Memorandum by the Minister for the Co-Ordination of Defence'. 19 March 1937.
40  CAB 23 42(37)7. 17 November 1937. For the proceedings of the Carbide Committee see CAB 16/174 'Committee of Imperial Defence. Sub-Committee on the Manufacture of Calcium Carbide in the United Kingdom'. July and August 1937.
41  CPRE 102/8. 'Caledonian Power. Petition' November 1937, 1–5.
42  *The Times*, 24 November 1937.
43  Ibid, 12 February 1938.
44  CAB 23 11(38)9. 9 March 1938.
45  CPRE 102/8. Handout. 14 March 1938.
46  *The Scotsman*, 23 March 1938.
47  P. Thomsen, *Scottish Water Power and in Particular the Caledonian Water Scheme* (Edinburgh, n.d., but probably, 1938), 38 and 40.
48  *The Caledonian Power Scheme* (APRS, 1938), 32.
49  'Water Power in Scotland: the Caledonian Scheme', *The Times*, 2 April 1938.
50  'The Glens and the Factory', *The Times*, 4 April 1938.
51  *The Times*, 5 April 1938.
52  Ibid.
53  Ibid, 6 April 1938.
54  *Hansard*, 334, 6 April 1938, 425 and 428.
55  Ibid, 435.
56  Ibid, 449.
57  Ibid.
58  Ibid, 447.
59  Ibid, 444.
60  Ibid, 464.
61  Ibid, 466.
62  Ibid, 467.
63  *The Times*, 7 April 1938.
64  'A Sorry Business', *E*, 15 April 1938, 474.
65  CAB 23 35(38)12. 27 June 1938.
66  See, in particular, *Hansard*, 342, 13 December 1938, 1781; 345, 4 April 1939, 2591–3; 346, 25 April 1939, 947–9; and 349, 4 July 1939, 1099–1101.
67  *Hansard*, 351, 21 September 1939, 1168.
68  Ibid, 361, 22 May, 1940, 148–9.
69  Ibid, 387, 25 May 1943, 199. See, also, the comments of Major Basil Neven-Spence, member for Orkney and Shetland, 229.
70  *Hansard*, 391, 28 July 1943, 1574–5.
71  Robert Boothby, *I Fight to Live* (1947), 55.
72  *Report of the Committee on Hydro-Electric Development in Scotland* (Cmd 6406, 1942). The report was savagely attacked by P. Thomsen in *The Cooper Report and the Hydro-Electric Development Bill* (Edinburgh, 1943).
73  *Hansard*, 387, 24 February 1943, 181.
74  Ibid, 193.
75  Ibid. Willie Gallagher was also highly critical of the constitutional status of the new board. *Hansard*, 389, 27 May 1943, 1798–1800.

76 Ibid, 194.
77 Leslie Hannah, *Engineers, Managers and Politicians: the First Fifteen Years of Nationalised Electricity Supply in Britain* (1982), 156.
78 Christopher Harvie, *No Gods and Precious Few Heroes: Scotland 1914–1980* (1981), 58.
79 Adam Collier, *The Crofting Problem* (1953), 166. As will be evident from the date of publication, this minor classic was issued posthumously.

# 8
# The Battersea controversy

If pro-electrical militants were convinced that opposition to the new source of energy would be restricted to the leisure class which inhabited England's broad and rolling acres – and Scotland's incomparable glens – they were disabused by the anti-'super-station' movement which flared, briefly but intensively, in London in 1929. Working people who had to endure the worst of the capital's fogs and smogs; professionals committed to the preservation of an environmentally privileged life-style; and an ex-Archbishop of Canterbury and King, who depicted large-scale emissions of sulphur dioxide as a threat to the social as well as the physical fabric of the nation: all stridently opposed the scheme to build a power station at Battersea. To those among the triumphalist lobby who had glimpsed the London Power Company's ambitious plans, Battersea represented a cathedral of electrical progress, whose massiveness and logistical centrality epitomised the imminent victory of electricity over every other form of energy. (Despite the crisis of 1929, it would, in due course, finally rationalise the capital's fragmented supply system.¹) It was inconceivable that Londoners should be intimidated by so magnificent and modern a building, or that they should ask embarrassing questions about the burning of 2000 tons of fuel a day. That some of them threatened to go to Parliament to obtain private injunctions against the construction of the station and equated the environmental repercussions of electrical technology with the 'primitive' coal and gas industries, was unthinkable. Yet the Battersea saga demonstrated that, in terms of 'amenity', the electrical *avant garde* was no more enlightened than its rivals. The blueprint made scarcely any provision for the prevention of atmospheric pollution: the boon of electricity was assumed to outweigh even the most pernicious side-effects.

The original application from the London Power Company to the Electricity Commissioners in October 1927, stated that 'the Company shall, in the construction and use of the said generating station, take the best known precautions for due consumption of smoke and for preventing as far as reasonably practicable the evolution of oxides of sulphur, and generally for preventing any nuisance arising from the generating station or from any operations thereat'.[2] But the sheer scale of the proposed plant and the fact that the site itself might be lost if rapid progress were not achieved during the existing parliamentary session, dictated that environmental measures were given a low priority. As for the Electricity Commissioners, they, too, were aware of the need to meet a burgeoning demand for electricity in the capital and to press on with the Grid. 'I may say for your private information', Sir John Snell, their chairman wrote to S. L. Pearce, engineer-in-chief at the LPC, 'that a memorandum I have just prepared on the Rate of Growth of Output in London satisfies me at any rate that it will be vital to proceed with Battersea and Chiswick simultaneously if there be no other difficulties from any objections which may be received.'[3] Parliament itself ruled out Chiswick as a suitable locale for a power station and, proceeding strictly according to statute, the LPC then invited complaints from households or businesses within a 300-yard radius of the proposed site at Battersea. Since there were only thirteen objectors, the LPC told the Commissioners that it might be possible to omit an inquiry and merely hold a hearing, 'so as to give interested parties an opportunity of ventilating their views in the matter'.[4] Snell opted for what amounted to an informal inquiry to be held at the Commissioners' headquarters at Savoy Hill. The government was kept fully in the picture and the Conservative Minister of Transport, Wilfred Ashley, was convinced that 'provided nothing very startling occurred ... the Commissioners would probably agree to it'. He was in any case ready to give the scheme his immediate consent if everything went according to plan.[5] Snell was optimistic as well, but he warned Ashley that although 'the Commissioners will almost certainly agree to give consent to the Battersea Station ... sometimes the unexpected occurs at these Inquiries, and delay may occur'.[6]

In the event the Savoy Hill 'inquiry' was a tepid affair. The number of complainants had now fallen to three and, only one, Mrs Handel Booth, threatened to raise the temperature: she believed that the company should be forced to pay compensation – 'England has not become so entirely bolshevised', she said, 'that individuals have not got a few rights'.[7] The LPC's counsel told residents that, since Battersea was already a predominantly industrial area, a power station would not lower the general environmental tone. Sweeping assurances were also given in relation to fumes directly attributable to the station. 'We have evidence', the small band of complainants were told, 'that there will be no emission here of gases which are objectionable.' Householders 'need have no apprehensions at all'.[8] If the Electricity Commissioners were now confident that they could press on and confirm the application, they had reckoned without the Office of Works, which was responsible for the protection of public and historic buildings in the capital. Stating that he was 'seriously concerned over the continual addition of further sources of atmospheric pollution in London', a senior official at the Office asked the Commissioners for an assurance that 'all reasonable precautions are taken to prevent the pollution of the atmosphere by smoke and other products of the combustion of coal'.[9] The Commissioners were unwilling to give the Office a legally binding undertaking that buildings would be exempt from damage – if they gave ground in this instance, numerous other interested parties in the capital might seek similar satisfaction. But, under pressure from the LPC, which was now eager to move from the planning to the construction stage, they agreed to a form of words which bound the company to 'prevent nuisance arising from the station and any operations thereat'.[10] For the moment, the Office of Works was satisfied; eighteen months later it, like many another public body and individual in London, would want to reconsider its position.

Although not carried on behind closed doors, the Battersea negotiations had thus far remained almost wholly insulated from the scrutiny of local political and environmental pressure groups. Residents of Chelsea, who prided themselves on the salubrity and desirability of their district, now joined the fray and sought assurances that fumes from the station would not seriously affect either stonework or health. 'In connection with the development of electricity supply', a deputation from the borough was told by Sir Harry Haward, the vice-chairman of the Commissioners, 'there was bound to be conflict between those interested in a cheap supply of electricity and those interested in the preservation of amenities, and the best that could be done was to arrive at a reasonable compromise between those necessarily opposing parties.'[11] Sensing that the delegation 'had based their objections on their experience of the [much smaller] Lot's Road Station', Sir Harry emphasised that Battersea would keep atmospheric pollution to a minimum and, to clinch his point, unveiled 'the latest steel gauge protection device' to filter out dust and grit. The Chelsea deputation was evidently less impressed than it should have been by this mechanism and asked whether it could be allowed to see the final plans for the station which would soon be agreed between the LPC and the Commissioners. Sir Harry attempted to delay, but left his visitors with the clear, though mistaken, impression that there would indeed be further, detailed negotiations.[12]

Two months later, at the beginning of 1928, borough opinion had hardened. The Town Clerk of Kensington informed the Commissioners that his council were 'far from satisfied' with the project and urged that it should be 'abandoned'. Fear of sulphur fumes in the capital was growing; but so, also, in the more affluent districts to the north of the river, was hostility towards a supposedly massive influx of building labourers needed to work on the station.[13] When more detailed plans became available in August, inter-borough animosity was added to existing concern about the environmental impact of the station in Chelsea, Kensington and Westminister. These councils now accused the LPC of coming to a secretive, unilateral agreement with Battersea Borough Council, and promising that it would adopt any reasonable anti-pollution precautions which 'the Council might from time to time suggest'.[14] The excluded boroughs demanded to be granted similar terms – whether, in the light of existing technical knowledge, such undertakings would actually have meant anything is a different matter – but the Commissioners refused. 'A further multiplication of safeguards through the medium of individual agreements with additional authorities' might 'have the result of incompatible requirements being made on the Company by different authorities.'[15]

Hostility and suspicion – between the 'excluded' councils, the LPC, the Electricity Commissioners, and Battersea Borough Council – were now intensifying in an atmosphere increasingly dominated by widespread public fear of the sulphur fumes. Early in December 1928, Sir Samuel Hoare, MP for Chelsea and Secretary of State for War, wrote to Snell expressing his concern and urging the Electricity Commission to use its influence to ensure that all affected districts should be given the same degree of legal protection. The Battersea affair, he added, might 'well become of importance to the general public whose anxiety as to the destruction of local amenities is more and more making itself felt'.[16] Snell replied at length, but concluded that there would be little point in his meeting the Mayor of Chelsea, as Hoare had suggested, since the matter had already been considered in depth. A week later, Reginald Blunt, the secretary of the environmentalist Chelsea Society, who would play a prominent role in subsequent events, complained bitterly of the Commissioners' refusal to bring about a meaningful agreement between the LPC and Chelsea Council.[17] Battle lines had been fixed.

'One thing we will guarantee', wrote an anonymous correspondent in the *Electrical Times* at the end of March 1929, 'and it is this – Wheresoever this power station be situate, that place shall be absolutely the worst, most objectionable, foolish, wicked and criminal. The prevailing wind will, in every case, blow directly upon the Nation's most sacred art treasures.'[18] This might well have been in reference to a joint letter published in *The Times* on 9 April, and signed by, among others, the Mayors of Chelsea and Westminster, the President of RIBA, the King's Physician, Lord Dawson of Penn, and representatives of a number of other preservationist and voluntary organisations. The statement focused, first, on the unprecedented size of Battersea and the extent to which its emissions would threaten the whole of 'historic' and 'institutional' London, including the Houses of Parliament, the National Gallery and Whitehall. It went on to catalogue the ways in which the construction of the superstation would exacerbate the capital's notorious fogs and thus pose a serious threat to public health. Surely German experience had demonstrated that vast undertakings of this type should be sited away from urban areas and close to coalfields? East Kent would be an ideal position, but if the LPC and the Commissioners refused to compromise, they must expect to be arraigned by large numbers of private injunctions and a formidable metropolitan protest movement.[19] *The Times* supported the collective letter and announced itself sceptical of any 'best practicable means' clauses designed to enforce higher standards on potential industrial polluters. The paper was disturbed by what smoke and dust might do to some of London's finest stonework and sympathetic to a remote, down-river site.[20] This point was reinforced by Reginald Blunt when he asked 'if Germany and other countries abroad can bring their electricity from the coalfields to their cities, why cannot we? If they can utilize the Rhine for cable transmission, why cannot we use the Thames? If buried cables are costly, has not the relative cheapness of pithead coal to be offset from *their cost?*'[21]

That the controversy was becoming increasingly intense was substantiated by the efforts of officials at the Ministry of Transport to ensure that both their own minister, and the premier, were more fully and convincingly briefed.[22] By the second week in April, King George V had become involved. Disturbed by *The Times* correspondence, he instructed his private secretary to write to Neville Chamberlain, the Minister of Health. The King was 'in entire sympathy with the views expressed by the signatories to the letter' and felt 'the greatest concern at the prospect of the atmosphere of London being still further polluted by the large quantity of noxious fumes which the station must inevitably emit'. 'Why', he wanted to know, 'should it not be possible to follow the example of foreign countries where power stations are erected at a considerable distance from the towns which they serve, and in surroundings where they do the least damage?' The King considered the London Power Company's project 'particularly ill-advised and hoped that the government will take steps before it is too late, to prevent it being carried out'. Chamberlain informed the King's secretary that, as Minister of Health, he had 'no authority to control the location of the generating station'.[23] Electricity was the responsibility of the Minister of Transport, but Ashley was not a member of the Cabinet; Chamberlain proposed, therefore, to circulate the letter containing the King's views to all senior ministers. When the issue had been debated in Cabinet he wrote to the Palace to report that 'we have received definite assurances that the emission of grit and smoke can and will be prevented'. As regards outpourings of sulphur dioxide, experts at the London Power Company had found a 'means of preventing this nuisance'. But there were problems of scale which had not yet been resolved, and the Laboratory of the Government Chemist as well as the Department of Scientific and Industrial Research would therefore be asked to monitor progress. 'Without an Act of Parliament', Chamberlain concluded, 'the Power Company could not be deprived of their legal right to erect a generating station capable of developing up to 120,000 kilo watts.' He was nevertheless ready to admit that it would in fact have been possible to construct the station on a down-river site – even though additional costs of transmission had been estimated at nearly a quarter of a million pounds a year.[24] But the King was unsatisfied with this response and his secretary wrote back to Chamberlain to insist that, if the process for the removal of sulphur dioxide could not be perfected 'the King trusts that the Government will suspend the working of the Station until such time as the results obtained prove entirely satisfactory'.[25]

Ashley, meanwhile, was trying to reassure the public, by means of a carefully prepared parliamentary reply, that the monitoring by the Government Chemist and the Department of Scientific and Industrial Research of developments at the LPC would lead to rapid and successful results.[26] Yet, if, following the King's intervention, the government was attempting to reduce widespread public anxiety, the Electricity Commissioners continued to follow a sturdily independent and unashamedly 'pro-electric' line. They informed Ashley that the economics of extra-urban siting placed such a policy beyond the realms of the feasible and that underwater cables in the Thames were impracticable and would, in any case, incur the wrath of the Port of London Authority. The Minister was advised to play down the possibility of wide-ranging environmental damage in London and to insist, in every public statement, that it was the LPC and not the government which was duty-bound, by statute, to put its house in order under the watchful and helpful eye of the appropriate scientific agencies. The Commissioners were eager to protect their own position and to insist that the LPC and not themselves must, in the final analysis, deal with any incipient public nuisance.[27] Now Lord Davidson of Lambeth, until recently Archbishop of Canterbury, added his voice to those of the protesters. Speaking from a holiday yacht in the Aegean, he telegraphed Chelsea Borough Council, which was still in outright opposition to the scheme, 'Please use my name in any effort against Battersea Power Station'.[28] Reginald Blunt also kept up the pressure and, in another letter to *The Times*, asked why the Commissioners had failed to insist on more effective anti-pollution measures, and doubted whether it would ever be possible to scale up the techniques which were currently being tested by the LPC.[29]

On 25 April the Battersea issue was debated in the Lords. An extreme opponent of the station, Lord Jessell, who was also playing a leading role in Westminster City Council's representations, deplored the omission of preparatory consultation with interested bodies in the capital, and underlined the absurdity of the '300-yard rule', which governed planning inquiries. There was no good reason, he insisted, why massive generating plants should not be sited far from the centres of large cities; and if the government failed to acknowledge the intensity of public anxiety it might be necessary to press for special 'disabling' legislation.[30] Lord Davidson, now returned from the Aegean, stated that he was not speaking 'on behalf of the residential population, in the sense of the better-to-do people so much as on behalf of the populace living around who ... are bound to suffer intensely from what must be an enormous production of what hitherto we have found to be both unpleasant and deleterious'.[31] Lord Birkenhead, for the government, proffered a somewhat casuistical defence of the '300-yard rule' and argued that the development of large-scale electricity supply, with its enormous employment-producing potential, must invariably outweigh every other social consideration.[32]

When Lord Londonderry, the First Commissioner of Works, sought to calm public fears, he was inhibited by the fact that, only three weeks earlier, his own department had decided temporarily to withdraw its support from the Battersea scheme. In a letter to Snell at the beginning of April, the Office of Works had complained that we have had before us what seems to be the best expert advice available, from which it appears not only that the new Power Station will cause a dangerously large concentration of sulphur dioxide in the air but that there are no methods in ordinary use which can prevent it ... The First Commissioner ... is dismayed to learn that the conditions attached to the consent of the Department are likely to prove practically worthless as a protection ... Is there any possibility of the scheme as a whole being reconsidered?[33]

All that Snell could offer by way of reassurance was that he was convinced that the LPC's research would dramatically reduce the discharge of smoke and grit.[34] Nor was this all, for on the very eve of the Lords debate, Londonderry had attended a meeting at which Chamberlain had reported that Dr Bailey, the Chief Alkali Inspector, 'knew of no means of preventing sulphur fumes, and doubted whether their prevention would be possible'. One of Bailey's colleagues had shared these doubts and 'feared also that it would not be practicable to reduce the emission of grit to unobjectionable levels'.[35] It was small wonder, then, that Londonderry's response lacked conviction and that, in the eyes of the Ministry of Transport and the Electricity Commissioners, it implied concessions which threatened both Battersea and the rapid construction of the Grid. Conceding that the first phase of the super-station was 'experimental', Londonderry went on to say that

one third [of the project] is being undertaken now, and this will have the result of eliminating three [environmentally harmful] stations already in existence. The other two-thirds will be undertaken later on, but before those two-thirds are undertaken there will be many opportunities for myself and the Government Departments to make representations. The matter will have to come under the purview of Parliament and I am quite ready to assure the most rev Lord that unless things are satisfactory the remaining two-thirds will not be undertaken.[36]

Londonderry's statement threatened the autonomy of each of the bodies – the Ministry of Transport, the Electricity Commissioners and the Central Electricity Board – concerned with national electrification. For, if it really were the case that every scheme which was believed to be technologically flawed, would be examined (and possibly re-examined) by Parliament and interested departments, the basis on which the construction and management of the Grid had been predicated – that 'business-like' agencies should press on, unharried by short-term swings in political and public opinion – would be undermined. Later in the year, Herbert Morrison, Labour's intensely pro-electric Minister of Transport, would withdraw the 'Londonderry concessions'. But, in the short term, environmental and borough pressure groups in London believed that the Minister had done too little rather than too much to reassure a worried public. In an article on the day after the debate, *The Times* complained that 'neither Lord Londonderry nor Lord Birkenhead appear to have realized the strength of the feeling by which they [the protesters] are actuated, and neither of them ventured to say that chemical research had yet reached the point where the emission of sulphurous fumes from the chimneys of power stations can be prevented'.[37] Nor did Londonderry's statement assuage the dissident boroughs intent on 'abandonment'. The Clerk to the City of Westminster Council insisted that the Electricity Commissioners should now 're-open the whole question of the erection of the proposed station', and unless the sulphur problem could be solved once and for all, ensure that the station be built on 'some suitable site outside the Metropolis where the emission of fumes, smoke, soot and dust cannot prove to be injurious to health and property'.[38] This demand was followed by an intervention by the influential Sir Richard Redmayne, a former Chief Inspector of Mines, which was critical of the role played by the Electricity Commissioners: their failure to search before 1927, for other and, in economic and logistical terms equally satisfactory sites; their unwillingness to warn the capital of the seriousness of the sulphur threat; and their determination, despite the current hiatus, to press ahead with a similar scheme at Fulham.[39] Nor were the authors of the original and influential joint letter to *The Times* satisfied with the government's latest reaction to the crisis. Re-evaluating the situation in the light of the Lords debate, they insisted that no guarantee had yet been given that the level of sulphur contamination could be contained. They attacked the argument put forward by the pro-Battersea lobby that a fully operational station would lessen pollution as a result of the reduced number of coal fires that would then be in use – domestic varieties of fuel, they contended, were much less noxious than the types usually burnt in power stations.

Finally, there was strong criticism of Londonderry's potentially misleading deployment of the term 'experiment' – what was being described here was not a controlled laboratory test, but the first part of a scheme, burning no less than 2000 tons of coal a day, and much more than that when the second and third sections were completed by the mid-1930s.[40] In another letter to *The Times*, Percy Lovell, speaking for the preservationist London Society, attacked Birkenhead's flimsy defence of the '300-yard rule' and pointed out that 'the ancient glass in the windows of York Minster is pitted by fumes attributed to the city of Leeds more than 20 miles away'.[41] The issues which were now being raised revolved around the constitutionality of public inquiry procedures, the accurate presentation of scientific information to lay audiences, and the *terra*

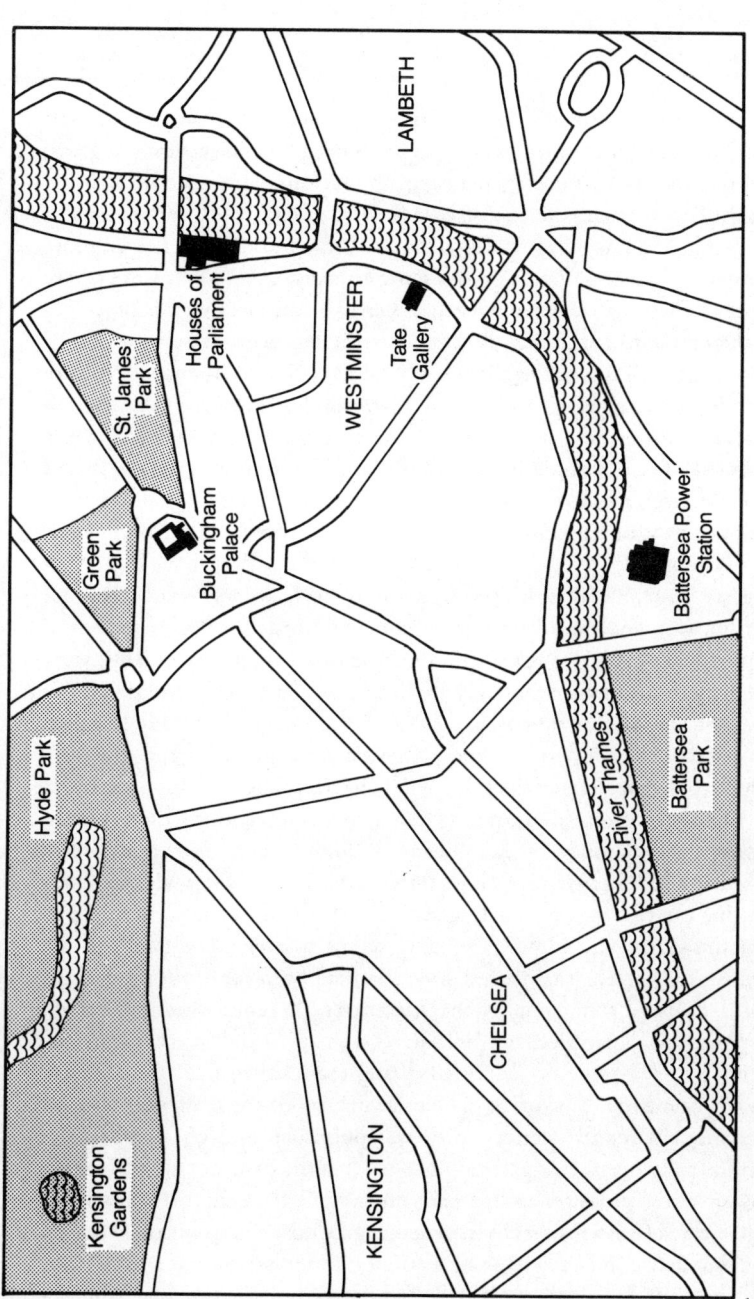

**Figure 10** Institutions and areas 'under threat' during the Battersea crisis

*incognita* of the long-term (and long-distance) impact of atmospheric pollution.

Linking the Battersea controversy to the anti-pylon movement, and contempt for 'humanists' who sought to hold back electrical progress, and national economic regeneration, the technical press now hit back. The intensely scientific *Electrical Industries and Investments* had insisted in April that 'if a few window-curtains have to be washed or renewed more frequently, we fear that window curtains must give way to electricity supply'.[42] The matter was evidently clinched by the fact that 'the objectors to the erection of the new power house at Battersea are non-technical men, for the most part, while those who have drawn up the scheme are experienced engineers'.[43] When the *Electrical Times* entered the debate it reminded its readers that 'posterity may find some amusement in the spectacle of a community which on the one hand clamours for cheap and abundant electricity, a boon which is promised by its leading politicians, and yet on the other hand fights tooth and nail to prevent the fulfilment of that promise'.[44]

The 'East Kent' solution was futile both in terms of costs of transmission and the adverse environmental impact of long stretches of high-tension cable. As for vociferous demands for 'abandonment', were the protesters unaware that only a special act of Parliament could stop the LPC? Surely 'all this ridiculous agitation' was in reality nothing less than an attempt to perpetuate 'the domestic coal grate ... and the rich sulphurous emissions of the gas fire'.[45] 'The power behind this opposition will naturally ignore all the points that do not suit it and will do its best to prolong the clamour by fair means or foul.'[46] The *Electrician* took a different and less paranoid view. Public criticism, it acknowledged, was 'getting out of hand; and, in that sense, to ignore the protest, in the hope that the storm will blow over, only serves to stiffen the backs of the opposition'. What was needed and what had been far too long delayed was a decisive statement by the Electricity Commissioners.[47] As if in response to this plea, an initiative was now taken, not by the Commissioners, but by W. F. Fladgate, the Chairman of the London Power Company. Surveying the entire history of the controversy in an exceptionally lengthy letter which was published in both the specialist and the daily press, Fladgate pointed out that the basic decision to have a power station at Battersea had been taken by Parliament in 1927 when it had rejected Chiswick on environmental grounds. This position had been unanimously confirmed by the Electricity Commissioners and both they, and the Central Electricity Board, had been emphatic that the Battersea scheme was integral to the Grid and to a rational supply system for the capital. The LPC had raised approximately £3 million on the open market, and contracts for two-thirds of that amount had already been signed, all of them involving British firms and British labour during a period of extreme economic instability. The widely expressed fear that the station would produce large quantities of life- and property-threatening sulphur dioxide was, in Fladgate's view, greatly exaggerated: no more than a vague extrapolation based on insecure and amateurish data.[48] The *Electrician* immediately reprimanded Fladgate for avoiding a crucial issue – a precise quantification of the amount of sulphur dioxide given off at each stage of production. 'Before', it commented, 'there is any talk of prior commitments, it should be clearly established, either that the sulphur problem is definitely solved or that the present scheme is economically better than the alternatives.'[49] But other sections of the trade press supported Fladgate's dismissal of the LPC's critics. 'We should like to see some qualitative and quantitative tests', the *Electrical Times* challenged the enemies of the super-station, 'with all the domestic chimneys in full blast and ... with only the factory chimneys at work. We believe that the contrast would silence a lot of this ridiculous talk as to the choking of babes and innocents of London with noxious fumes. Such a test could quite easily be made.'[50] 'Gross exaggeration of the amount of sulphurous fumes ... is very rife amongst the critics. People who ought to know better are talking of "many hundred tons" of sulphurous oxide discharged daily from the new station at Battersea. However, violent language of this kind has its benefits. When a saucepan boils over too much it puts the gas out.'[51] Supporting the view that the protesters had greatly overstated the dangers of sulphur fumes, *Electrical Industries and Investments* concluded that 'the number of people who know nothing about the technicalities and problems concerned who have "barged in" is astonishing'.[52]

By the end of June 1929, the new Labour Minister of Transport, Herbert Morrison, had drawn up a policy which he hoped would reduce public disquiet and minimise the occurrence of embarrassing disputes and misunderstandings between government departments: the Ministries of Transport and of Health, as well as the Office of Works would liaise on a regular basis in an effort to solve the Battersea problem. Morrison, reacting against the Londonderry concessions, was determined to stand firm against 'any action which might delay or prejudice the bringing into operation of this scheme [and which] might have serious consequences including an adverse effect on employment'.[53] If ground were given over

Battersea, there might be a rash of injunctions against power stations in every part of the country. This could not be tolerated.[14]

But the government was not yet out of the woods. The London County Council, which had come to a satisfactory agreement with the LPC following the inquiry at Savoy Hill in 1927, was now distancing itself from the dissident boroughs; but Westminster, Chelsea and Kensington remained committed to the struggle and Westminster, in particular, was attempting to co-ordinate yet another anti-Battersea front and arguing that it was 'far better for the local authorities concerned to pay such an indemnity [up to £2 million] than to be saddled with a power station that might cause continuous injury to health and damage to property'.[15] It was only slowly dawning on the militants that the first phase of the station might actually be completed before 'disabling' legislation could be brought before Parliament.[16] The tide was now turning and, by the end of July, Fladgate felt sufficiently confident to issue a press statement which announced that 'the Power Company desire to state in the most emphatic manner that their Engineers and Chemists have evolved a perfectly practicable and commercial solution of the "sulphur problem" in which they have complete confidence, and that the menace feared will be non-existent so far as the Battersea Station is concerned'.[17] But protest continued, and in October, Morrison and Arthur Greenwood, the Minister of Health, decided that it might be useful to talk to a delegation from the London County Council. The two Labour ministers were, of course, among friends – Morrison, in particular, had close connections with London politics and politicians – but they also hoped that the LCC representatives would use their influence to dissuade the dissident boroughs from taking any further action against the Battersea scheme. Greenwood pointed out that 'operations had ... already been begun at Battersea and [there could be] no question of stopping the work that had been put in hand'. Morrison placed the issue in a wider context. Any proposed anti-Battersea legislation would need to contain significant compensation clauses in favour of the LPC. As for the suggestion that there might be some kind of 'freeze' on the construction of further stations, that 'would be to say that the supply of electricity in London was to be retained at its present level for five or six years, which could have serious repercussions upon employment, domestic amenity and business development'.[18] The meeting between Morrison, Greenwood and the LCC delegation was a cordial one but 'anti-electric' individuals refused to accept that the LPC could really solve its technical difficulties, or that the new government would prove any more successful in dealing with the super-station dilemma than the Conservatives. *The Times* remained sceptical. 'Failing', it stated, 'the fulfilment of that condition [successful control of sulphur dioxide emissions], the Government, as the only body with power to prevent the damage have a very serious responsibility to London and its inhabitants, to say nothing of the rest of the Empire of which London is the capital.'[59]

Lord Dawson of Penn, echoing the King, attacked the LPC with undiminished ferocity. 'That the nation', he thundered,

> should be asked to permit a gamble over the fabrics of Westminister Abbey and the Houses of Parliament, over damage to our priceless collections of pictures in the National and Tate Galleries, and over the trees, grass, and flowers in our parks, let alone the health of the people is unthinkable. I suggest that a consensus of expert opinion, supported by several years' experiment in other districts would alone justify the further prosecution of the Battersea scheme.[60]

Fladgate responded immediately. 'I think that he [Dawson] owes an apology . . . for interfering in a matter of this importance without taking the trouble to make himself acquainted with the facts of the case.'[61] But Dawson would not be silenced. 'Having been scourged with whips', he asked in biblical mood, was not the community 'justified in making certain that the London Power Company will not, through error of judgement, scourge it with the scorpions of sulphur oxides and acids?'[62]

By the end of October Morrison had received further chemical results from the LPC. 'The report of the Government Experts is satisfactory so far as it goes', he told his colleagues, 'but leaves it open to the opponents of the scheme to press again for the abandonment of the Battersea Station or, at any rate, for a cessation of work, until the further experiments can be concluded.'[63] The Government Chemist stated that the research undertaken thus far indicated that 'it is possible to eliminate from the gases of combustion nearly the whole of the sulphur gas present'. But he was not yet convinced that scaling-up would be totally successful and felt that 'a more definite explanation of the oxidation of sulphurous to sulphuric acid . . . should be forthcoming before we can report on the practicality of the process'.[64] 'It looks as if the officials concerned have been three and a half months making up their minds – to be cautious', the *Electrical Times* sceptically commented when the results were published.[65] But the LPC team, ably led by Pearce, was clearly on the right lines, and by the beginning of November 1930, Morrison was welcoming a further report which confirmed that 'with an ample capacity in the gas-washing plant . . . affording thus a sufficient time of contact, and with iron as a catalyst, theelimination of sulphur from the gases, without any use of soda or other alkali, should reach 80 to 85 per cent at least . . . Generally, we may say that the Company has achieved the object of eliminating sulphur gas from the flue gases on a fairly large scale.'[66]

The battle of Battersea Power Station, which had briefly united a large number of interest groups and individuals who believed that chemical devastation would be spread by a plant which had yet to be built, was nearly over. Early in that year of confrontation – 1929 – Henry Tizard at the Department of Scientific and Industrial Research had privately confessed that 'there is a strong prima facie case in support of the contentions of the objectors to the proposed Battersea Station . . . We believe that there is no doubt that the effect of such high concentrations [of sulphur dioxide] will be injurious to vegetation and buildings.' Any clause insisting that the LPC should agree to take anti-pollution measures which were 'reasonably practicable' was, in his view, valueless since there was no known agent which could remove the delinquent substances in a manner that could in fact be described as 'practicable'. 'It does seem a little odd to an amateur', Tizard concluded, 'that a large power station should be put down in the middle of London . . . I did hope that one of the great advantages of the grid scheme would be the possibility of avoiding what is now happening.'[67]

The real, 'objective' facts about the station – how large a volume of deleterious gas and dust the unmodified plant would have produced; how technically effective the LPC's solution to the problem would prove to be; and how significant a contribution fumes from Battersea continued to make to the fogs and smogs which bedevilled the capital until the 1950s: these can never be fully recovered or reconstructed. But a number of points may be made by way of conclusion. The first relates to Tizard's belief, which was shared by large numbers of non-scientists, that superstations should never be sited – did not in fact 'belong' – in large and long-established urban centres. Communal fear may, in that respect, have been as closely related to what Mary Douglas has termed 'matter out of place', as to any 'real', measurable threat to the environment.[68] Secondly, there is the question of presentation and legitimation. So long as ill-informed ministers stumbled from crisis to crisis and reassurance to reassurance. Those who opposed the building of the station were repeatedly able to marshal credible and convincing counter-arguments. But, when scientific agencies were formally delegated to monitor the LPC's efforts to reduce pollution, the protesters' arguments began to lose their persuasiveness. This had less to do with the ability of scientists to solve problems relating to sulphur emission than with the increasingly widespread belief that it was the specific social *role* of scientists to solve such problems. Had the anti-Battersea militants been more sceptical towards scientists and 'official science' and better informed about the record of government and industry in dealing with atmospheric pollution in other spheres, they might have fought harder and longer.[69]

Finally, there is the question of property. The King and other members of the social élite who opposed the Battersea scheme did not, of course, dismiss the impact of sulphur dioxide on the 'health of the people'. But they clearly considered it to be much less important than what the fumes might do to palaces, works of art, parks and gardens. To aristocratic and upper-middle-class preservationists, the metropolis possessed an overwhelmingly important ornamental and symbolic status – it was, after all, still widely recognised and revered as the 'heart of the Empire'. As such, it constituted a socially exclusive environment, an idealised *polis*, in which ordinary men, women and children merely happened to live, work and have their being.[70]

### Notes

1 For a survey of electrical systems in the capital up to 1930 see Thomas Hughes, *Networks of Power*, chap.9.
2 PRO POWE 12/140. London Power Company letter of application. 27 October 1927.
3 POWE 12/140. Sir John Snell to S. L. Pearce. 30 March 1927.
4 POWE 12/140. W. W. Lackie to the Electricity Commissioners. 24 May 1927.
5 POWE 12/140. Memorandum from Wilfred Ashley to Snell. 17 June 1927.
6 POWE 12/140. Snell to Ashley. 17 June 1927.
7 POWE 12/141. Public Inquiry transcript. 21 June 1927, 76.
8 Ibid, 24.
9 POWE 12/140. F. J. E. Raby to Electricity Commissioners. 20 July 1927.
10 POWE 12/140. Secretary of Electricity Commissioners to H. M. Office of Works. 27 July 1927.
11 POWE 12/161. Memorandum of Interview. 10 November 1927.
12 POWE 12/161. Correspondence between Town Clerk of Chelsea Borough Council and the Electricity Commissioners. 12 and 16 November 1927.
13 POWE 12/161. Town Clerk of Kensington Royal Borough Council to the Electricity Commissioners. 24 January 1928.
14 POWE 12/161 'Copy of agreement between the London Power Company and Battersea Borough Council, dated April 27, 1927.'
15 POWE 12/161. Assistant Secretary of Electricity Commissioners to the Town Clerks of Westminster and Chelsea Borough Councils. 2 October 1928.
16 POWE 12/161. Sir Samuel Hoare to Sir John Snell. 4 December 1928.

17  POWE 12/161. Snell to Hoare (n.d.).
18  'Opposition to the New Battersea Station', *ET*, 28 March 1929, 471.
19  *The Times*, 9 April, 1929.
20  Ibid, 'A Power Station in Battersea'.
21  *The Times*, 11 April, 1929.
22  POWE 12/231. Sir Cyril Hurcomb to Sir John Brooke. 12 April 1929.
23  POWE 12/140. Captain A. M. L. Hardinge to Neville Chamberlain. 12 April 1929. Chamberlain replied on 15 April and again on 18 April.
24  POWE 12/140. Chamberlain to Hardinge. 15 April 1929. The Cabinet deliberations are contained in CAB 23/60 17(29)11, 17 April 1929. For further background see CAB 24/202/114.
25  POWE 12/140. Hardinge to Chamberlain. 21 April 1929.
26  CAB 24/203/115. 'Battersea Power Station: Memorandum by the Minister of Transport.' April 1929.
27  Ibid. 'Electricity Commission. Battersea Power Station. Memorandum by the Commissioners', 1–5. 15 April 1929.
28  *The Times*, 16 April 1929. The current Archbishop of Canterbury, Cosmo Lang, was also drawn into the dispute. See Rob Cochrane, *Landmark of London: The Story of Battersea Power Station* (CEGB, n.d., probably 1984).
29  Ibid, 23 April 1929.
30  *Hansard*, House of Lords, 74, 25 April 1929, 210–15.
31  Ibid, 223.
32  Ibid, 217–19.
33  POWE 12/140. Sir Lionel Earle to Sir John Snell. 2 April 1929.
34  POWE 12/231. Snell to Earle. 16 April 1929.
35  POWE 12/140. 'Record of a Meeting . . . ' 24 April 1929.
36  *Hansard*, House of Lords, 74, 25 April 1929, 225–6.
37  *The Times*, 'Battersea Power Station', 26 April 1929.
38  POWE 12/231. Clerk to City of Westminster to Secretary, Electricity Commission. 26 April 1929.
39  *The Times*, 30 April 1929.
40  Ibid, 1 May 1929.
41  Ibid, 3 May 1929.
42  *EII*, 4 April 1929, 588.
43  Ibid, 'Pros and Cons of Battersea', 8 May 1929, 75.
44  'Opposition to Super-stations and Overhead Lines', *ET*, 9 May 1929, 711.
45  Ibid, 'The Agitation against the New Battersea Power House', 717–18.
46  'Agitation Against Battersea Power Station', *ET*, 16 May 1929, 757.
47  'Battersea Again', *E*, 10 May 1929, 548.
48  W. F. Fladgate, 'The Battersea Power Station', *E*, 17 May 1929, 577–8.
49  'The Battersea Project', *E*, 24 May 1929, 605 and Ibid, 'An Alternative Site', 605.
50  'Who Fills the London Air with Fumes?', *ET*, 30 May 1929, 835–6.
51  'The "Sulphurous Fumes" Campaign', *ET*, 6 June 1929, 877.
52  *EII*, 26 June 1926, 1067.
53  CAB 24/202/177. 'Battersea Power Station: Memorandum by the Minister of Transport', 3. 25 June 1929.
54  Ibid, 11–12.

55   *The Times*, 25 July 1929. The 'dissident' boroughs had held a meeting at Westminster City Hall on 2 July. See *ET*, 18 July 1929, 87.
56   On the threat of parliamentary action see POWE 12/231. Clerk to the City of Westminster to Sir John Brooke. 27 July 1929; and on doubts about its feasibility, *The Times*, 26 July 1929.
57   POWE 12/231. Copy of letter by W. F. Fladgate. 26 July 1929.
58   POWE 12/231. 'Battersea Power Station. LCC Deputation'. 15 October 1929.
59   'Battersea Power Station', *The Times*, 16 October 1929.
60   Ibid, 19 October 1929.
61   Ibid, 21 October 1929.
62   Ibid, 25 October 1929.
63   CAB 24/206/293. 'Battersea Power Station: Memorandum by the Minister of Transport', 2. 29 October 1929.
64   'Treatment of Sulphur Fumes in connection with the Working of the proposed Electric Power Station of the London Power Company at Battersea' (HMSO, 1929), 9–10.
65   'Sulphurous Fumes Report', *ET*, 12 December 1929, 985.
66   'Treatment of Sulphur Fumes in connection with the Working of the Proposed Electric Power Station of the London Power Company at Battersea: Second Report' (HMSO, 1930), 4.
67   POWE 12/140. Tizard to Geoffrey Fry. 11 February 1929.
68   Mary Douglas, *Purity and Danger*, 35.
69   For background on attempts to control atmospheric pollution see R. M. MacLeod, 'The Alkali Acts Administration, 1863–84: the Emergence of the Civil Scientist', *Victorian Studies*, IX, 1965, 85–112 and Anthony S. Wohl, *Endangered Lives: Public Health in Victorian Britain* (1983), chap. 8.
70   I have tried to give an account of 'imperial' symbolism in nineteenth-century London in my *Pollution and Control: a Social History of the Thames in the Nineteenth Century* (Bristol, 1986), chap. 1.

# 9
# Arcadia under threat

It is now time to probe more deeply into why it was that the individuals and pressure groups described in earlier chapters so sternly resisted the march of the pylons. This will involve an analysis of the ideologies underlying environmentalism and preservationism in inter-war Britain, an evaluation of the connections between these bodies of ideas and other modes of thought, and a more comprehensive account of social and political interests. In order to become better acquainted with this underexplored terrain, we shall be scrutinising the writings and agendas of four important activists and propagandists – G. M. Trevelyan, Vaughan Cornish, Clough Williams Ellis and Patrick Abercrombie. This 'common core' of environmental ideas should provide us with a guide to the collective consciousness of rank-and-file opponents of the Grid and of urban super-stations.

The great-nephew of Lord Macaulay, and Regius Professor of History at Cambridge from 1927 to 1940, G. M. Trevelyan was the best-known and most popular historian of his day. He made his reputation with the Garibaldi trilogy and studies of politics under the Stuarts and Queen Anne, but then gravitated towards social history. During his retirement Trevelyan wrote the immensely successful *English Social History: a Survey of Six Centuries: Chaucer to Queen Victoria* (1942). A passionate lover of the English countryside, Trevelyan was President of the Youth Hostels Association from 1930 to 1950 and a prominent member of the National Trust. He died in 1962.[1] Compared with Trevelyan, Vaughan Cornish was an intensely private and, in worldly terms, unsuccessful man. Born in Suffolk in 1862, he was educated at Owens College, Manchester and worked briefly as Director of Technical Education in Hampshire. Supported by a substantial private income, Cornish was able to resign in 1895 and devote the rest of his life to free-lance geographical and scientific research. He published widely on travel, preservationism and the 'science of natural scenery', and spoke on behalf of numerous conservationist organisations, including the CPRE. Without honour during his own lifetime, Cornish died in 1948, but has recently been reclaimed by the academic community as a founding father of the controlled study of the perception and aesthetics of the natural landscape.[2]

Now probably best known as the creator of the neo-Venetian tourist trap of Portmeirion in North Wales, Clough Williams Ellis was a man of many parts in preservationist and planning circles in inter-war Britain. Educated at Cambridge, where he read science, Williams Ellis qualified as an architect but then gave as much time to writing and preservationist propaganda as the design of the revivalist and anti-modernist buildings of which he was a staunch and lifelong advocate. His adaptations and additions to Stowe School and his neo-baroque chapel at Bishop's Stortford – the first British building to be officially listed – have worn well. But it is as the author of *England and the Octopus*, published in 1928, and an intensely active founder-member of the Council for the Preservation of Rural England and its sister-organisation for Wales, that Williams Ellis is most likely to be remembered. He died in 1978 at the age of 95[3]. Patrick Abercrombie, a close friend of Williams Ellis, also started life as an architect but, by the outbreak of the First World War, he had established himself as a leading academic exponent of town planning at the University of Liverpool. During the inter-war years a flood of influential regional plans – for Deeside, the East Kent Coalfield, the Thames Valley and for Cumberland – flowed from his pen. In 1935 Abercrombie moved to the chair of town planning at University College, London and, during the war, produced the formidable *County of London* and *Greater London* plans. He was for many years an influential policy-maker for the CPRE. Abercrombie died in 1957.[4]

From his very earliest days, Trevelyan was deeply committed to, and indeed obsessed by, the relationships between man and nature. In his own life this took the form of exceptionally long and punishing walks near the family estate of Wallington in Northumberland. In his youth Trevelyan could manage forty miles a day and still as an old man he would attempt to recreate these early tramps 'with the help of a car and a chauffeur whom one could send round from point to point'.[5] Walking represented an escape from the claustrophobia of town life and the anguish of mental labour. As the walker slogged relentlessly on, driving himself to the very limit, the mind became no more than an instrument 'to register the goodness and harmony of things'. As for the body, it was transformed into 'an animated part of the earth we trod': 'drugged with sheer health', it was experienced 'only as a part of the physical nature that surrounds it and to which it is indeed akin'.[6] For Trevelyan, this love of and total physical and at times masochistic immersion in natural beauty seemed to predate all 'systems' and 'dogmas'. It represented 'the highest common denomination [sic] in the spiritual life' not just for the educationally and aesthetically awakened but for everyone.[7] In an increasingly urbanised world – a world 'denaturalised' by the activities of scientists[8] – every man and woman must occasionally escape for refreshment to the untouched 'uplands'.[9] 'Our modern life', Trevelyan wrote in 1929, 'requires such days of "anti-worry" ', and these could only be obtained 'in perfection' when 'the body has been walked to a standstill'.[10] These are clearly the words of a writer and intellectual. Trevelyan was a man whose intermittent depressions could be shifted by massively demanding physical exertion. (He actually walked many of the military routes described in his histories of Italy.) He believed urban culture – characterised more by the mechanistic 'detective story' or 'problem novel' than by the poetry which he loved so passionately[11] – to be essentially anti-humanistic; and he was ready, like other preservationists in inter-war Britain, to project his own, highly idiosyncratic vision of idealised relationships between man, woman and nature on to the great mass of city-dwellers who now visited the countryside by bus, car, bicycle and on foot. The history of England – but not, it should be noted, the history of *Britain* – was of overriding importance here. For Trevelyan, however, a specific national past was predated or presaged by a more generalised and frankly utopian vision. He touched on this primordial set of human relations when he wrote that 'it is indeed in the depths of natural wilderness that a man feels most united to his ancestors, for there he is for a moment withdrawn from the present noisy age, left alone with nature as his fathers were left alone amid the same green sights and quiet sounds'.[12] But this arcadia required and received a more specific identification and location. It became 'Trevelyan's England', a perfect rural environment, studded with village communities and peopled by classless individuals who were uniquely well-attuned to the aesthetics of unadulterated nature. (They even, so Trevelyan said, 'appreciated the sound of bird-song more intensely than any other race'.[13])

But Trevelyan's arcadia degenerated into an ugly, machine-ridden and utility-dominated nation of factories and 'profit'. First, 'the chess-board predictability' of enclosure tamed and ironed out the 'uplands', where men's souls could be refreshed.[14] Even more pernicious in terms of the way in which it broke the 'line of nature' was the nineteenth-century railway system.'15 For Trevelyan, the repercussions of this 'law of the machine age' were traumatic. 'By the end of the Nineteenth Century', he wrote, 'it was already true that almost all that was old was beautiful, almost all that was new was either ugly in itself, or in shrieking disharmony with the natural beauty amid which it was set.'16 As intensely as Toynbee or the Hammonds, Trevelyan inveighed against the collapse of the 'natural order' which, in his view, was indissolubly linked to the 'rationalisation' of agriculture following enclosure and the destruction of the domestic system through the introduction of the factory.17 Whether Trevelyan really 'believed' in this catastrophic fall from grace – one suspects that he did, with all his heart – is less significant here than the terms in which he described so total an environmental and cultural deterioration. The overriding impression, as with other preservationists who were deeply concerned about the pylon and super-station issues, is that he was as preoccupied with the collapse of the patrician social order as with any explicitly technological threat to an 'unspoilt' rural terrain. In the late 1920s and early 1930s, the landed élite was under serious economic threat. Land was being bought up by speculators and developers and these, according to Trevelyan, were groups which had little interest in the preservation of either the 'village community' or the environment in which it was located. Embryonic planning controls did, of course, exist, but the 'old' rural governing class was too often unwilling to fight its corner.18 'So much', Trevelyan urged in 1929, a year of intense 'anti-electric' activity, 'depends on men and women of the right sort being elected to serve on [local authorities] and consenting to stand for election.'19 Eight years later, when he contributed to Clough Williams Ellis's controversial *Britain and the Beast*, Trevelyan was still chiding the 'jerry-builder' and the 'exploiter' and the state's failure to protect amenity and the rural social order.20 Here, more vividly than anywhere else, we are confronted by the historian's individualism and conservative anti-statism. 'The State is Socialist enough', he wrote, 'to destroy by taxation the classes that used to preserve rural community; but it is still too Conservative to interfere in the purposes to which land is put by speculators.'21

Trevelyan, as a preservationist, seemed to write and act out of intuition, lending his support to whichever movement or pressure group might help to restore a rural arcadia. The geographer Vaughan Cornish attempted a more academic exercise: the systematisation and popularisation of an objective 'science of scenery'. A firmly rooted theory of environmental aesthetics could have greatly strengthened the hand of those who protested against the intrusion of large-scale electrical technology into rural – and urban – areas in Britain between the wars. Indeed, the absence of clear-cut academic or quasi-scientific criteria that 'landscape A' should be preserved at the expense of 'landscape B' had gravely weakened the environmental lobby and made it vulnerable to the charge that, in the final analysis, choices and options in this sphere were wholly subjective. Such accusations could be even more damaging whenever it could also be implied that 'landscape A' was being aggressively defended because it belonged, or was aesthetically attractive to an élite group. But, Vaughan Cornish only rarely engaged with the problems inherent in such an enterprise. His writings tend to be concerned less with the possibility of making a claim for absolute 'scientific' neutrality in this sphere than the necessity of empathising with the natural world and using each and every sense to experience it to the full. Cornish also insists, in a manner reminiscent of Trevelyan, that periodic bouts of isolation in the wilderness are the only sure means of temporarily healing the wounds attributable to urban civilisation. Finally, and crucially, he argues that experience of the natural world is the only incontrovertibly *true* and authentic experience – his 'science of scenery' is in fact predominently deployed to cast doubt on the validity and social utility of revealed religion.

The starting-point, as one might expect, is with Wordsworth, 'Here by Grasmere', Cornish wrote in typically florid style in 1937, 'Wordsworth kindled the torch which first lit up the temple of Nature for our worship. Then Ruskin by the shore of Coniston kept the flame alight, and to these, more perhaps than to any other prophets of the century now past, we owe the faith, now beginning to be manifest, that the beauties of nature are a source of inspiration unclouded by intellectual error and available for all men.'[12] Wordsworth was also claimed as a 'theoretical' ancestor. Cornish found in the poet 'scientific originality, for he not only records physical appearances, but also wherever they give keen enjoyment, seeks the source of the impression, investigating both the objective conditions and the mental qualities concerned in their appreciation'.[13] In the best of his writings Cornish achieved precisely that mix of immediacy and reflexivity which he so admired in his mentor. Elsewhere, though, he sought something different, though no less venerable in terms of its cultural lineage: a psychological state in which he would himself no longer be separable from nature. 'The breaking of the waves upon the shore, the heaving of the waters out beyond, and the tang of the salt sea breeze, so

## Arcadia under threat

took possession of all faculties and senses that I lost myself, and my being seemed to pass into that of the elements.'²⁴ 'The word death came clearly before me as an expression devoid of meaning, and so complete was this impression that no thought of a future state came within my ken. The conception of time had broken down and in cosmic union a moment was snatched from eternity.'²⁵ This was the Cornish who desired and advocated absolute solitude in a 'world made one by Immanence Divine'.²⁶ His explicitly religio-aesthetic preservationism was also, like Trevelyan's, deeply patriotic. He never forgot, and indeed grossly sentimentalised the quality of rural life in Debenham, Suffolk, where he had spent his childhood.²⁷ He idealised 'Arcadian England with its coppice woodland and deciduous trees, and fields divided by a bushy fence with hedgerow timber'.²⁸ Where for Trevelyan the primordial landscape had been the mountain range, Cornish was committed to a quite different arcadia – 'that which displays the life of ordered agriculture'.²⁹ Here clustered 'the villages of old-world architecture in which English scenery is unrivalled',³⁰ and it was in these places, before the invention of the combustion engine, that the main street had been 'a place of pleasant loitering and friendly gossip'.³¹

Trevelyan was convinced that the English country-dweller was uniquely attuned to bird-song; Cornish believed that 'Englishmen have an unusually keen perception of the beauty of trees', 'in all that relates to the beauty of trees the ordinary Englishman has the artist's eye'.³² He placed special emphasis on the aesthetic splendour and inherent healthfulness of the sea. 'Let it never be forgotten', he warned, 'that the sea view from the cliffs is the special scenic heritage of our island people'.³³ But all this was now threatened by the indiscriminate construction of a ramshackle 'seaside', consisting of beach-huts, ice-cream kiosks and litter-bins.³⁴ Cornish probably possessed a greater empathy with the needs of the urban working class than Trevelyan, but he still insisted that natural beauty must never be sacrificed to 'mere utility', and that the uninitiated must be educated in rural ways before they could be granted full access to wilderness or seascape.³⁵ Cornish, like Trevelyan again, was deeply pessimistic about the 'urban condition' and his solutions were radical. 'The project for slum clearance', he wrote in 1937, 'and that for national parks ought to be envisaged together as complementary parts of one great movement for saving England from what is mildly termed undue urbanization, a condition that is to say in which towns are not fit to live in and the countryside not fit to look at'.³⁶ An enthusiast of the garden city movement, Cornish advocated massive investment in urban high-rise development, thus allowing 'almost all the added space' to become 'available for town gardening and afforestation'.[37] Suburbia was anathema. It ate inexorably into the countryside via ribbon development; it led to 'soullessness' and 'lack of character' and it had shown itself to be 'lamentably destructive of the civic sense'.[38] Precisely why the suburbs should have been so vitriolically castigated is not clear, but, like many inter-war environmentalists, Cornish tended to think and write in terms of opposites. The real characteristics and 'morality' of suburbia eluded him.

Despite his commitment to an objective 'science of scenery', Vaughan Cornish devoted most of his working life in the inter-war period to a minute examination and depiction of a national landscape which he believed to be threatened on every side. The best of his writings were good and were taken up and imitated by lesser talents who were as firmly convinced as the Edwardians had been that the city and city-dwellers would undermine civilised national values. Seen from this perspective, Cornish's romanticism – and likewise Trevelyan's commitment to a mystical view of the relationship between man and 'wilderness' – was as much reaction to economic recession and the 'urban condition' as to a genuine threat to the countryside. The anciently established rural élite was in retreat before the perceived onslaught of developers and speculators and it was for this, among other reasons, that both Cornish and Trevelyan deployed the rhetoric of patriotism and arcadian traditionalism in a belated effort to restore the patriciate to something of its former glory.

Clough Williams Ellis shared many of these values. He savagely attacked developers, speculators, and the planners and builders of what he believed to be the most monstrous of inter-war seaside communities, Peacehaven in Sussex. Where Trevelyan and Cornish bemoaned the desecration of the uplands and the coastline, the Welsh architect and designer produced an instant manifesto which specified the environmental 'state of the nation' and popularised a rhetoric and vocabulary which would dominate debate for the ensuing decade. *England and the Octopus* established Williams Ellis as the impresario of the preservationist movement. Like Trevelyan, Cornish and his friend Abercrombie, Williams Ellis made use of an analogy between the rampant *laissez-faire* of the peak period of industrialisation in the mid nineteenth century and the free-for-all in the building and development booms of the 1930s.[39] He believed, with Cornish, that the great 'rush' to live and gain relaxation in the countryside was a condemnation of the alienation of urban existence and a dire threat to the survival of rural beauty.⁴⁰ Like Cornish, again, he was convinced that the slump could be creatively used to restructure great urban regions and bring them into closer conformity with the ideals of the garden city movement. But a major difficulty here was that the garden city ethos was itself only fully understood by a minority of the educated public. 'Welwyn is so scandalously unique', Williams Ellis lamented, 'that few even of those who have heard of it really grasp what it is, what it has done and what it stands for.'⁴¹ In places such as these, he believed, inhabitants would experience 'the charm, seemliness and essential rusticity which the poor bungalow dwellers try to get, and of which their own ignorance and the rapacity of the speculator so cruelly cheat them'.⁴² Williams Ellis truly detested speculators – this is a splenetic theme which dominates *England and the Octopus* and which threatens to undermine its creative vigour – but he was no less angry with politicans and their failure to incorporate 'amenity', 'environment' and 'orderly design' into party manifestos.⁴³

What was needed was a campaign of preservationist education allied to the 'co-operation of the privileged'.⁴⁴ 'Most people', Williams Ellis insisted, 'are neither with nor against us, merely from lack of imagination or lack of thought or observation, and their indifference and inertia we must try to overcome by effective propaganda.'⁴⁵ The body which would co-ordinate this educative programme – the Council for the Preservation of Rural England (together with its sister-organisations in Wales and Scotland) – had been founded three years earlier. The CPRE, he argued, 'must enlist the people of light and leading into its support, both men and women, especially such fortunate persons as are in such a position that they can prove their faith by works'.⁴⁶ If activity of this type were not immediately undertaken, the country would become 'a museum in which are preserved here and there carefully selected and ticketed specimens of what England *was*. The National Trust is England's executor.'⁴⁷ At the very end of *England and the Octopus*, Williams Ellis summarised those aspects of the natural and built environment most urgently in need of orderly planning by central government. His wrath fell especially heavily on ribbon development, petrol stations and electrical transmission lines. 'Disfiguring little buildings', he wrote, 'grow up and multiply like nettles along a drain, like lice upon a tape-worm: we may be rude about them like that – but we do nothing to check their increase.'⁴⁸ 'Might it not be the case', he went on, that the country would be 'so overburdened with a plethora of pumps that petrol-selling ceases to be remunerative, resulting in the disfiguring of our highways by derelict stations, even more unsightly in their dilapidated abandonment than in their flouting youth?'⁴⁹ As for transmission lines, they crossed 'the loveliest little valleys of Merioneth, destroying their scale and imparting a sense of sophistication and "Progress" which they were till now so soothingly innocent.'⁵⁰

Two years later Williams Ellis had become even more pessimistic: and 'order', 'cohesion' and a sense of the past were now central to his critique of 'anti-planning'. 'We are making a sad mess of our country', he lamented, 'with very little sense of order, seemliness, dignity or even efficiency, and none at all of that discipline without which there can be no assured or permanent freedom."⁵¹ There must be a central plan 'governing the use and treatment of the land that we call ours, but which was our fathers' and shall be our children's and is therefore ours only as trustees'.⁵² His dominant imagery was of the absolute cultural superiority of a countryside facing destruction at the hands of industrialists, ribbon developers, cars and ignorant day-trippers. If it wished to preserve the 'essential England', government must commit itself to a national conservation scheme, enforced by a Ministry of Amenity. Advertisement hoardings would be taxed at high levels in every place of natural or architectural beauty; industrialists would be required to make payments into a 'special reserve fund' to pay for activities leading to pollution, or for failures to clean up a site at the end of a designated phase of construction; arterial roads would be replaced by more 'natural' American-style parkways; and there would be significant national investment in education for conservationism and 'citizenry'.⁵³

Clough Williams Ellis did not himself contribute to *Britain and the Beast* in 1937 but his choice of essayists indicates profound disillusion with governmental failure to protect either 'wilderness' or the integrity of the built environment. Keynes, Forster, Trevelyan and Abercrombie provided incisive pieces, but other writers manipulated the theme of rural crisis to present blatantly anti-working-class and anti-urban polemic. These authors are quoted less for the merit or logic of their views than for what they reveal of what was often implicit in the writings of Trevelyan, Vaughan Cornish and Williams Ellis: fear of the city and nostalgia for a vanishing social order. C. E. M. Joad recommended without irony that habitual offenders against country 'lore' should be imprisoned.⁵⁴ Another contributor wrote that 'we the readers and writers of this book, and our like, are that [preservationist] minority. What shall we teach them and how get the lesson learnt? . . . They make up the England of today, watching bad films, listening to bad music, reading bad literature.'⁵⁵ Overwhelmingly patronising contempt for the 'ordinary' day-tripper reached a zenith in the essay by G. C. Hines on 'Cultural Pilgrimage'. 'The tragedy', he wrote,

is that their eyes, once blinded are always blind. That is why the annual charabanc outings of these poor maimed folk are occasions for every sort of offence against the beauty that England has to offer them. To some the litter-strewing drink-and-card-party in the pinewoods is a source of jest, to others of wrath; but wrath should rather be directed against those damnable conditions that have crippled and emasculated the aesthetic man, and if there be any laughter, it should be close to tears.[16]

Patrick Abercrombie subscribed to many of the ideas which have been identified as the stock-in-trade of preservationist ideology in Britain between the wars. He believed that 'the most essential thing which *is* England is the Countryside, the Market Town, the Village, the Hedgerow Trees, the Lanes, the Copses, the Streams and the Farmsteads'.[17] England, he was convinced, had 'invented the village and she has certainly perfected her invention; nor is there anything like it in any other country'.[18] But this arcadian serenity was now under threat. 'A single bungalow roofed with pink artificial tiles, a factory chimney at a focal point or a glaring advertisement is able at a stroke to destroy the composed beauty ... of a landscape."[19] Advertisements were especially pernicious. Numerous garages, Abercrombie complained, were 'faced with tin petrol advertisements which do not even scruple at the vilest degradation of the national flag'.[60] Such sentiments merged at times into a total condemnation of all that was modern and mass-produced and, on these occasions, Abercrombie's tone was remarkably similar to that of his friend and fellow-founder of the CPRE and CPRW, Clough Williams Ellis. One extraordinary diatribe inveighed against 'gramophones, roundabouts, cock-shies, automatic sweet machines, tea booths, picture postcards, touts and crippled mendicity'.[61] Abercrombie claimed that his experiences as a planner had taught him that 'untidiness' in town and country was 'extremely infectious: a few years of dirty and squalid industrialism and the evil effects spread over a whole countryside'.[62]

Many of his writings were tinged, also, with anti-urbanism and, even as late as 1937, when he had become distinctly cool towards purist preservationism, he could still write that it was a misfortune to 'have to live and work in towns and suburbs'.[63] But despite all this, Abercrombie – 'our greatest town-planner', according to W. A. Robson [64] – possessed a keen understanding of novel interactions between town and country in inter-war Britain. His major intellectual debts were to Leplay and Geddes, and throughout his career as academic and planner he continued to place great emphasis on the analytic 'triad' of 'place', 'man' and 'occupation'.[65] The notions of 'community' and of the *interdependence* of communities were also central to his thinking. There was little point, he insisted, in believing that villages could continue to flourish in isolation from the culture of the town or city. 'Every region', he wrote in classically Geddesian style, 'must have its metropolis in which the higher functions of Education, Art, Business and Pleasure are placed.'[66] Changing transport patterns and a hypothesised new potential for rural industry triggered by the extension of electricity had convinced Abercrombie that the protection of the agrarian past for its own sake was neither viable nor desirable.[67] 'Bluntly speaking', he wrote, 'we want to see if it is possible to put a great many more buildings, new roads, subdivisions of property etc., into the countryside and yet preserve its beauty either substantially as it is or in a changed form.'[68] These ideas were rooted in a belief that even the wildest and most remote landscapes were partly man-made – the product of explicitly social and historical processes. According to this view, it was possible for 'man' and 'nature' to coexist in a dynamic and aesthetically satisfactory dialectic. The increasingly complex blending of the 'natural' and the 'technological' was more than a historical inevitability: it was a process to be encouraged by those who wished to retain all that was most enduring in rural culture.[69] There was little comfort here for supporters of what Abercrombie termed the 'Local Materials' solution. This was predicated, he said, on the 'fallacy that the English countryside is a Museum piece that can only be preserved and repaired as we treat a ruined abbey, and that all the character and individuality must be eliminated from any additions'.[70]

Abercrombie gave more sustained attention to the issue of access than Trevelyan, Cornish, or Williams Ellis. Genuinely wild country, he insisted, must be left absolutely wild, and access guaranteed by a government committed to a national parks policy.[71] Elsewhere, and particularly in the great arable heartland, walkers must not expect to be allowed to wander at will.[72] By the time that he made his contribution to *Britain and the Beast* in 1937 Abercrombie had become sceptical of governmental commitment to planning for amenity. He castigated the lack of co-ordination between the Ministries of Health, Transport, Agriculture and the Board of Trade and pointed to a continuing lack of control over ribbon development. In the sphere of architecture and design, he warned that 'the country is in danger of losing beautiful old cottages and gaining ugly new ones'.[73] On the eve of the Second World War Abercrombie was openly impatient with government and its failure to play a dominant role in town and country planning. Re-emphasising his commitment to the social philosophy of Geddes and Leplay, and insisting that it was now necessary 'actively to construct the *environment* in which the human *organism* may rightly *function*', he put forward a programme which included land nationalisation and municipal ownership.[74] The war would provide Abercrombie with an opportunity to produce his most influential reports – the *County of London Plan* and the *Greater London Plan* – but, by the later 1930s, he was experiencing deep disappointment both with his former allies in the rural preservation movement, and with a government which still refused to grant sufficient powers either to local authority agencies or to the planning profession itself.

Abercrombie's perceptiveness as social scientist and planner, and his awareness of dynamic interactions between town and country in inter-war Britain, enabled him partially to break with the regressive romanticism which so often shackled Trevelyan, Cornish and Clough Williams Ellis. Their style of thought, with its insistence on a uniquely English arcadia, a rural landscape polluted by the excrescence of 'development', and the need for mass education in countryside appreciation and 'citizenry', can be tangentially related to Martin Weiner's recent work on the 'decline of the industrial spirit' in twentieth-century Britain.[75] But, in terms of an explanation of the ideological structuring and chronological development of rural preservationism something more specific is needed than Weiner's insistence on the seductiveness of the rural way of life and the sets of symbols and myths that are intermingled with it. Two traditions were strongly at work in the writing and thinking of preservationists during these years: a 'nature mysticism', reaching back to early-nineteenth-century romanticism but retaining its appeal, in an increasingly conservative form throughout the Victorian period; and a barely suppressed antagonism towards the urban working class and urbanism in general. Something has already been said about that ideology and its interconnections with preservationism; all that need be added here is that, with the by now predictable exception of Patrick Abercrombie, each of our writers and activists was confused by the traumatic urban decline which had taken place in the north of England, lowland Scotland and North Wales, during these years. Despite an insistent advocacy of the garden suburb ideal, each was, as we have seen, intensely suspicious of 'faceless' suburbia. Suburbs were thought to be inimical to a sense of community and 'civic spirit' and to encourage everything that was most reprehensible in terms of 'anti-planning'. But another and possibly more powerful factor was also at work. Suburbia was by definition neither wholly urban nor wholly rural and it was for this reason that its growing cultural importance disturbed a polarity which had been central to British and European social thought for more than three centuries.[76] The near-eradication of that polarity was confirmed by the growing numbers of town-dwellers and suburbanites who now unapologetically followed leisure pursuits in rural areas. This was an invasion which was perceived as disrupting both 'wilderness' and the insulated tidiness of the 'domesticated' rural heartland and as further undermining the system of social relations by which the traditional order had been underwritten.

What, finally, should be said about the rural élite in inter-war Britain? Analysis of the writings of G. M. Trevelyan, Vaughan Cornish, Clough Williams Ellis and Patrick Abercrombie strongly suggests that the instability and fragmentation of the social order in the countryside in these years was rooted in objective socio-economic decline. The intensity of the agricultural depression; the impact of death duties; unprecedentedly high levels of sales of land at freehold; and the growing influence of families who had made their fortunes wholly outside the agricultural community – these processes were the primary shapers of rural preservationism.[77] A specific sense of social insecurity was frequently and unambiguously sublimated into the protection of an arcadia under threat.

## Notes

1 On Trevelyan see the autobiographical works cited below and the *Dictionary of National Biography*.

2 For recent reassessment of Vaughan Cornish see E. W. Gilbert, 'Vaughan Cornish (1862–1948) and the Beauty of Scenery' in his *British Pioneers in Geography* (Newton Abbot, 1972), 227–56 and Andrew Goudie, 'Vaughan Cornish – Geographer', *Transactions of the Institute of British Geographers*, 55, 1972, 1–16.

3 There is a useful entry on Williams Ellis in the *DNB* and additional information in two autobiographical works. *Architect Errant* (1971) and *Around the World in Ninety Years* (1978).

4 For concise biographical material on Abercrombie see Gerald Dix, 'Patrick Abercrombie 1879–1957' in Gordon E. Cherry (ed.), *Pioneers in British Planning* (1981), 103–30.

5 G. M. Trevelyan, 'Autobiography of a Historian' in *An Autobiography and Other Essays* (1949), 43.

6 G. M. Trevelyan, 'Walking' in *Clio: a Muse and Other Essays* (1929), 5 and 16.

7 G. M. Trevelyan, 'The Call and Claims of Natural Beauty', in *An Autobiography and Other Essays*, 105–6.

8 Ibid, 101.

9 Ibid, 95.
10 'Walking', 17.
11 'The Calls and Claims of Natural Beauty', 94.
12 G. M. Trevelyan, *Must England's Beauty Perish? A Plea on Behalf of the National Trust for Places of Historic Interest or Natural Beauty* (1929), 13.
13 Ibid, 20. Trevelyan's vision was finally encapsulated in the massively successful *English Social History: a Survey of Six Centuries: Chaucer to Queen Victoria* (1942).
14 'The Calls and Claims of Natural Beauty', 102.
15 'Autobiography of a Historian', 14–16.
16 'The Calls and Claims of Natural Beauty', 104.
17 There is a strong, though indirect, link here with D. H. Lawrence. See, for example, 'Nottingham and the Mining Countryside' in Edward D. McDonald (ed), *Phoenix: the Posthumous Papers of D. H. Lawrence* (1936), 133–40. Trevelyan's pessimistic preservationism, like Lawrence's abhorrence of the 'machine', may have been rooted in trauma associated with mass destruction during the First World War. For suggestive comment on this point see John Sheail, *Rural Conservation in Inter-War Britain* (Oxford, 1981), 7.
18 For concise accounts of planning mechanisms at this time see Sheail *passim* and John Stevenson, *British Society 1914–45* (1984), 221–43.
19 G. M. Trevelyan, *Must England's Beauty Perish?*, 23.
20 G. M. Trevelyan, 'Amenities and the State' in Clough Williams Ellis (ed.), *Britain and the Beast* (1937), 186.
21 Ibid, 183.
22 Vaughan Cornish, *The Preservation of our Scenery: Essays and Addresses* (1937), 83.
23 Vaughan Cornish, *National Parks and the Heritage of Scenery* (1930), 83.
24 Vaughan Cornish, *The Poetic Impression of Natural Scenery* (1931), 20.
25 Ibid, 9.
26 Ibid, 44.
27 See note 23 above and Vaughan Cornish, *National Parks and the Heritage of Scenery*, 71–3.
28 *The Poetic Impression of Natural Scenery*, 4.
29 *The Preservation of our Scenery*, 45.
30 Ibid, 8.
31 Ibid, 36.
32 *National Parks and the Heritage of Scenery*, 46 and *The Preservation of our Scenery*, 18.
33 Ibid, 57.
34 Ibid, 33–4.
35 *National Parks and the Heritage of Scenery*, 45.
36 *The Preservation of our Scenery*, 71.
37 Ibid, 31.
38 Ibid, 74 and 45–7.
39 Clough Williams Ellis, *England and the Octopus* (1928), 24–7.
40 Ibid, 38.
41 Ibid, 44–5.
42 Ibid, 48.

43  Ibid, 70–9.
44  Ibid, 91.
45  Ibid, 97.
46  Ibid, 104.
47  Ibid, 108.
48  Ibid, 162.
49  Ibid, 168.
50  Ibid, 144.
51  Clough Williams Ellis, 'The Modern Landscape' in 'H.H.P.' and 'N.L.C.' (eds), *The Face of the Land* (1930), 12.
52  Ibid, 16.
53  Ibid, 20–4.
54  C. E. M. Joad, 'The People's Claim' in Clough Williams Ellis (ed.), *Britain and the Beast* (1937), 80.
55  Geoffrey M. Boumphrey, 'Shall the Towns Kill or Save the Countryside?' in *Britain and the Beast*, 103.
56  G. C. Hines, 'Cultural Pilgrimage' in *Britain and the Beast*, 161.
57  Patrick Abercrombie, *The Preservation of Rural England: the Control of Development by means of Rural Planning* (Liverpool, 1926), 6.
58  *The Preservation of Rural England*, 14.
59  Ibid, 16.
60  Ibid, 27.
61  Patrick Abercrombie, *Country Planning and Landscape Design* (Liverpool, 1934), 22.
62  Patrick Abercrombie and others, *The Coal Crisis and the Future: a Study of Social Disorders and their Treatment* (1926), 39.
63  Patrick Abercrombie, 'Country Planning' in Clough Williams Ellis (ed.), *Britain and the Beast*, 133.
64  W. A. Robson (ed.), *The Political Quarterly in the Thirties* (1971), 28.
65  *The Preservation of Rural England*, 55.
66  *The Coal Crisis and the Future*, 37.
67  *The Preservation of Rural England*, 7 and *Country Planning and Landscape Design*, 19.
68  *The Preservation of Rural England*, 50.
69  *Country Planning and Landscape Design*, 5–6.
70  Ibid, 27.
71  *The Preservation of Rural England*, 16.
72  Patrick Abercrombie, *Town and Country Planning* (1933), 217.
73  Patrick Abercrombie, 'Country Planning' in *Britain and the Beast*, 139.
74  Patrick Abercrombie, *Planning in Town and Country: Difficulties and Possibilities* (Liverpool and London, 1937), 53.
75  Martin J. Weiner, *English Culture and the Decline of the Industrial Spirit* (Cambridge, 1981).
76  The literature on this theme is large and diffuse but a stimulating interpretation is provided in Raymond Williams, *The Country and the City* (1973). See, also, Andrew Lees, *Cities Perceived: Urban Society in European and American Thought 1820–1940* (Manchester, 1985).
77  See, on this theme, F. M. L. Thompson, *English Landed Society in the*

*Nineteenth Century* (1963), chap. 12; Lawrence Stone and Jeanne C. Fawtier Stone, *An Open Elite? England 1540–1880* (Oxford, 1984), part IV; David Lowenthal and Hugh C. Prince, 'The English Landscape', *Geographical Review*, 54, 1964, 325–31; and John Stevenson, *British Society 1914–45*, 332–6. Edith M. Whetham provides an overview of developments in inter-war agriculture in *Agrarian History of England and Wales VIII, 1914–1939* (Cambridge, 1978).

It was only after I had finished the final draft of this chapter that I derived the full benefit of reading the brilliant depiction of 1930s ruralism contained in Valentine Cunningham's *British Writers of the Thirties* (Oxford, 1988), chap. 7. This account is in many ways similar to my own but has the advantage of dealing with a larger number of writers and propagandists and working with a more widely focused ideological lens. It cannot be too strongly recommended.

10

# Nuclear aftermath

Between 1945 and 1950 the nationalisation of the industry ensured that electricity established itself in a growing number of urban areas and penetrated more deeply into hitherto poorly served country districts. As electricity and science education spread, wonder and fear declined. But some of the old problems – or variants of them – persisted. In 1987, more than half a century after the completion of the Grid, no fewer than 40 people were reported to have been the victims of electrocution, and in the same year 170 deaths were directly attributed to fires related to the use of electrical appliances. Six per cent of the total adult population suffered a non-fatal shock; four out of ten did not know if or when their home had been rewired; and a third admitted that they undertook every form of repair, however complex, without the aid of a professional electrician. John Butcher, the Conservative Minister for Consumer Affairs, reminded the public, in tones reminiscent of Caroline Haslett, that 'electricity is a powerful force that we all tend to take for granted, but it is a force that can kill, maim and burn'.[1] A rush of reports which hypothesised connections between the clustering of high-tension wires and increased incidence of cancer and depression, were uncannily reminiscent of ideas that had been laughed out of court by electrical progressives in the 1920s.[2]

Triumphalism, meanwhile, had lost none of its persuasiveness or cultural ubiquity. Deploying terminologies which echoed the social, technocratic and nationalistic assumptions of electrical enthusiasts in the aftermath of the Great War, the seminal White Paper *A Programme of Nuclear Power* insisted in 1955 that 'nuclear energy is the energy of the future'.[3] The survival of Britain as a major international power was argued to depend 'both on the ability of our scientists to discover the secrets of nature and on our speed in applying the new techniques that science places in our grasp'.[4] This abstraction of 'science' from its social context had permeated the ideology and rhetoric of triumphalism in the inter-war period: mastery of the universe was then expressed in power stations and the National Grid. Now, the government contended, it was 'only by coming to grips with the problems of the design and building of nuclear plant that British industry will acquire the experience necessary for the full exploitation of this new technology'.[5] Such an enterprise was replete with historical and imperial as well as economic and industrial associations – Britain was urged to fulfil its 'traditional role as an exporter of skill, to the benefit both of ourselves and of the rest of the world'.[6] It was confidently predicted that that most troubling of national domestic dilemmas – the Janus-like 'coal problem' which had preoccupied politicians and technocrats for more than thirty years – would be solved by the application of the new and socially neutral source of energy. 'The provision of enough men for the mines', the White Paper stated, 'is one of our most intractable problems and is likely to remain so ... The mining industry will ... remain one of the major employing industries of the country, but it may hope to be relieved by the advent of nuclear power of the excessive strains which are now being put upon it.'[7]

As for popular fears of nuclear power – shaped and sharpened by the horrors of Hiroshima and Nagasaki and the constant threat of full-scale atomic conflict between America and the USSR during an arctic phase of the Cold War – these were to be thrust aside in the interests of a national scientific and technological revolution. 'If nuclear facilities are properly designed any accidents that may occur will be no more dangerous than accidents in many other industries ... The reactors that will be built for the commercial production of electricity will present no more danger to people living nearby than many existing industrial works that are sited within built-up areas. Nevertheless, the first stations, even though they will be of inherently safe design, will not be built in heavily built-up areas.'[8] British electrical progressives during the inter-war years had given comparably sweeping and unsupported reassurances to 'non-believers'; now an important section of the bureaucratic and policy-making apparatus was backing a scientific élite which was ill-informed or over-sanguine about many of the constructional problems, as well as the massive environmental threats, associated with nuclear energy. Far more than the 1930s, this was an era in which scientists and technologists were insulated from public opinion and in which government presented idealised 'power' as a national panacea. 'Our civilization', the White Paper concluded, 'is based on power'[9]; thus it was that 'this formidable task must be tackled with vigour and imagination. We must keep ourselves in the forefront of the development of nuclear power so that we can play our proper part in harnessing this new form of energy for the benefit of mankind.'[10]

Shielded from governmental, academic and lay criticism, and legitimated by appeals to Britain's historical and imperial 'mission', the nuclear establishment was granted *carte blanche* to proceed with large-scale technological and environmental experimentation. As in the 1920s and 1930s, so now in the mid-1950s, Parliament remained docile in the face of triumphalism. In February 1955 the Conservative Ronald Bell initiated a poorly attended debate which amounted to little more than a formal vote of thanks to the Atomic Energy Authority and the drafters of the White Paper. 'In the sphere of nuclear development', Bell enthused, 'Britain has run right ahead of the rest of the world.'[11] Bell was also convinced that the first generation of stations would be 'inherently safe, because if the moderator inside them does get overheated, if the process begins to go too fast, it stops itself '.[12] As for public apprehension about nuclear waste, this was 'entirely fantastic'.[13] Echoing Bell, Bernard Braine stated that 'atomic energy has given our country a new lease of life',[14] a sentiment which was uncritically taken up by normally sceptical Labour members like Fred Willey and Arthur Palmer. Beating a nationalistic drum, Willey insisted that 'we must see that in this initial stage of the new world that is opening before us Britain is in the lead'.[15] For the government, Nigel Birch, the Minister of Works, provided the by now mandatory salute to the 'new nuclear age' and reiterated that it was essential on economic and nationalistic grounds for British scientists and British expertise to be at the forefront of the atomic revolution. He was adamant, also, that the peaceful deployment of the new form of energy must be disassociated from pervasive images of mass slaughter, and public anxiety about the dangers of nuclear waste refuted.[16] Outside the citadels of science and government, nuclear triumphalism was more sceptically received. When, at the beginning of 1956, Brian Harrison, the Conservative MP for Maldon, took informal soundings at Bradwell-on-Sea in Essex, which was to be one of the two initial sites, he reported a wide range of attitudes. Although he estimated that the safety issue had been 'worrying local residents more than anything', Harrison identified others like Tom Driberg and John Betjeman, whose main aim was to ensure that the village would be 'preserved as a show piece'. Farmers feared that constructional work would deprive them of scarce labour in an underpopulated corner of the county, while other members of the community were convinced that the Central Electricity Authority would make excessive demands on water supplies for cooling purposes and deface the landscape with its massive lines of pylons.[17]

Like his predecessors in the inter-war period, Aubrey Jones, who was Minister of Fuel, hoped that it might be possible to bypass a public inquiry.[18] But opponents of the scheme had already noted contradictions in *A Programme of Nuclear Power*. If, they argued, the new energy source posed safety threats which were no more serious than those encountered in conventional industrial activities, why must places like Bradwell be forced to pay so high an environmental price: surely the stations should be constructed in urban areas and thus gain the economic benefits of being located at the centre of major communications networks. Jones's officials realised that there was no convincing answer to this critique – the White Paper had failed to square the circle on the safety issue – and advised him to fall back on the rather feeble rationale that the siting had been 'determined only in part by "safety" considerations'.[19] When Jones received a delegation from Bradwell at the end of February safety loomed large. So also did criticism of the venerable '300-yard rule'. How was it possible, Driberg and his colleagues demanded, for the minister to weigh public opinion in so far-flung a community if he allowed himself to be hamstrung by a procedure which gave a statutory hearing solely to those living or having an economic interest in property within the immediate physical vicinity of the proposed site?[20] Any embarrassment which the minister experienced during the meeting was counterbalanced by the news, during the first week of March, that at an informal parish gathering, large numbers of villagers had declared themselves to be in favour of the scheme.[21] Was this yet another example, sceptics asked, of an articulate rural élite, claiming to speak for the 'community', but acting in such a way as to deprive that 'community' of the fruits of technological progress? One thing was certain. Militant opponents of the siting of the station were clearly better organised, and better able to influence public opinion than those like Tom Driberg's gardener who was in favour of the scheme because he believed that it would 'probably mean a better school for my children'.[22] The villagers constituted a silent, pro-nuclear majority, but, by the beginning of the second week of March, Jones had received more than a 150 formal objections – eventually there would be over 500 – and therefore had little option but to sanction a public inquiry.[23]

The hearing opened in Bradwell Village Hall at the end of April. The first witness was the unswervingly triumphalist Deputy Director of the Engineering and Industrial Group of the Atomic Energy Authority, P. T. Fletcher. Echoing the White Paper, Fletcher insisted that 'if nuclear power facilities are properly designed, any accidents that may occur will be no more dangerous than accidents in many other industries'.[24] In response to the question, 'Will the kind of effluent that you make from this plant be capable of being used, so far as its radioactivity is concerned, as drinking water without the slightest danger?', he answered 'Yes, that is perfectly true'.[25] The Deputy Director was also convinced that 'the amount of radioactivity which even at the most pessimistic estimate one could contemplate being lost from a plant of this type is so small that it should be quite readily possible to warn persons in the immediate neighbourhood of the hazard . . . and to arrange for appropriate advice to be given'.[26] Had he been subjected to cross-examination by counsel with even a minimal grasp of the technical issues at stake, Fletcher could hardly have appeared so sanguine.

J. D. Peattie, the Acting Chief Engineer for the Central Electricity Authority, who had clearly been briefed to volunteer as little information as possible, was more harshly treated. Why, in the light of what had been said about safety in *A Programme of Nuclear Power*, did the CEA continue to insist on a rural rather than an urban site? Why had proposed construction schedules and completion dates not yet been made public? Why had so little attention been given to the logistical difficulties of transporting components for the station round the narrow, high-hedged roads which led into Bradwell village? And why was Peattie unwilling to commit himself over the possibility of a life-threatening malfunction?[27] In response to each of these questions, and many others, Peattie stonewalled; the format, as well as the political and ideological weighting of the public inquiry system allowed counsel to probe so far, but no further.

When Tom Driberg took the stand, it became clear that 'amenity' and 'environment', construed in their pre-war, preservationist senses would predominate over technical considerations and the fear of nuclear power and nuclear waste. Driberg complained that the CEA had failed to make vital technical information available to the protesters, but he was also adamant that he did not wish to 'overstress the safety aspect, because personally I am not afraid of its going off or anything'.[28] Here he deployed images and assumptions which were *lingua franca* among both defenders and opponents of the peaceful uses of the new form of energy: a nuclear accident was still conceived predominantly in terms of a military nuclear *explosion*. When he inveighed against the environmental impact of the proposed development, Driberg's presuppositions and phraseology were uncannily similar to those used by Vaughan Cornish and Clough Williams Ellis. 'It is perfectly possible', he told the inquiry, 'that other industrial plants may be built in the neighbourhood, and that each such development seems to bring with it inevitably a sort of rash of hideousness – of shacks, shanties, cafes and notice-boards – which ruin the countryside almost more than anything and which are creating universally in England ... a kind of "Subtopia", which is neither true town nor true country.'²⁹ The modernity of the power station would clash grotesquely with the antiquity of the marshland environment: 'a building of that sort rising out of our subtly delicate horizontal landscape cannot but be an eyesore'.³⁰ All this came from a *soi-disant* epicurean socialist who had claimed, during the scientific days of the construction of the Grid, that 'objectively looked at, a pylon is as graceful as a steeple and can only give scale to a landscape'.³¹

Like Driberg, the spokesman for the Council for the Preservation of Rural England adopted the vocabulary and social perspective of an older style of conservationism. His words are worth quoting at length as an illustration of the underlying ethos of 'traditional' post-war environmentalism; and as an indication of how little was known, outside restricted scientific circles, of the health and safety implications of the 'new nuclear age'. The typical visitor to Bradwell was

consciously or unconsciously trying to escape ... when he turns his back on urban and industrial life and seeks mental and spiritual refreshment in these wide open and uncluttered spaces of land and sea and sky. I think it is true to say hitherto this strangely haunting and attractive country has found its symbol in the modest, lovely and ancient chapel of St Peter in the Wall, and it will, in our view, be a sad day for Bradwell and all it stands for when that symbol is superceded by the dominating bulk of the Bradwell Nuclear Power Station.³²

In his summary J. B. Herbert, who had so ruthlessly cross-examined Peattie, laid major emphasis on the arcadian tranquillity of the marshes. 'There are many people', he said, 'to whom the very name "Bradwell" brings on emotion as powerful as that brought by the name of, say, Oxford to some people, or Romney Marshes to others. And it is a misapprehension on the part of the Authority if they think, as apparently they did think, that they can ignore the existence of that strong emotion.'³³ Communal anxiety over possible nuclear accident was presented as being far less intense than that associated with 'the mass opening up of the district to the day-trippers [which would] drive away the sort of people who have enabled Bradwell people to make a living ... the yachtsman who once used to patronise the village may go ... [and] there will be substituted for them a mob of strangers with no interest in the place, in the country, or in the scenery at all'.³⁴

'Industrialism' would despoil the landscape and disrupt the local economy. 'I fear', Herbert announced in the tones of Cornish and Williams Ellis, 'the semi-urbanisation of the area.'[35] 'With more traders coming in from outside ... more houses will be required, more garages and petrol pumps will be erected, and more flotsam and jetsam of industrialisation will come with them.'[36] (Even now, in the mid-1950s, the lowly petrol pump retained it potency as a symbol of philistine 'modernisation' and disruption from without.) 'It is', Herbert concluded, 'as much a crime to put this power station on the estuary of the Blackwater as it would be to put a soap factory in Kew Gardens.'[37]

Little of this was lost on the mandarins at the Atomic Energy Authority who were pressing Aubrey Jones for a rapid and positive decision. 'It seems', wrote D. E. H. Peirson, the secretary of the AEA, 'that the main weight of the opposition is concerned with factors quite unconnected with the nuclear nature of the proposed power station.' The principal objectors, he went on, were the amenity lobby and the fishing interest which was convinced that over-heated waste water from the station, together with excessive chlorination, would destroy the Blackwater's famous oyster-beds. But these problems, Peirson insisted, were not directly related to nuclear energy *per se*; and if Jones felt that he had no option but to reject the CEA's application 'on grounds unrelated to the nuclear nature of the prososal, this latter circumstance should be made clear in the Minister's statement. Obviously, rejection of the Bradwell application will be a very heavy blow to the nuclear power programme; but it would be very much more serious if an unfounded impression were to be created that the Minister had withheld his assent because of the nuclear nature of the project.'[38] In the event, Jones, who was under heavy political pressure to keep to the demanding timetable set out in *A Programme of Nuclear Power*, accepted his inspector's report without qualification. The site had been found to be technically sound. In environmental terms, the inspector reported that there was a discrepancy between the intensity of protest and the number of people actually visiting and making use of Bradwell and its immediate surroundings. The complaints of the fishing interests were rejected on the grounds that large natural fluctuations in the oyster population outweighed any possible detrimental impact of an increase in the average temperature in the estuary. Local water supplies were also argued to be likely to be enhanced rather than threatened and the poor quality of the country roads judged to be no more than a minor hindrance to the delivery of components.[39]

As in the South Downs in 1929, so now, in Essex in 1956, the public inquiry system had been deployed to neutralise local preservationist opposition and boost triumphalist claims for a new and world-transforming technology. There were, as we have already seen, strong elements of continuity – the long-established domination of the environmentalist camp by upper- middle-class protectors of an aracadian order in a new nuclear age; the probable existence of a silent, working-class majority, convinced that atomic power would bring an ill-defined social liberation to their community; and a minister under heavy pressure from the scientific élite, the electrical industry and 'progressive' public opinion. But there were significant differences. Inspectors at public inquiries might still do little to mask the fact that they were institutionally and professionally biased towards the aims of government and public corporations: evidence was patiently heard but when it ran wholly counter to the dominant ideology of triumphalism it tended to be ruled out of court or judged to be 'outweighed' by countervailing factors which reinforced the aims and needs of the state. But the burgeoning complexity of organised science, and the increasingly threatening environmental implications of technological progress were already beginning to reshape the quasi-judicial discourse which characterised these forums. During the inter-war years debate had centred on the visual and 'aesthetic' harm which would be done to landscapes if they were criss-crossed by lines of pylons. These preoccupations, which continued to mediate a powerfully 'English' preservationist ideology, were still dominant at Bradwell. But, running alongside and beneath them – nagging sub-texts, frequently played down, joked about, or ignored by the inspector – were explicitly scientific, medical and ecological concerns. Already in 1956, between discussion of rural 'aesthetics', the protection of ancient monuments, and the right of the rambler to continue to enjoy unlimited access to 'wilderness', counsel intermittently cross-questioned experts, and quoted specialist literature on water pollution, oysters and ornithology.

Pre-war preservationist ideology retained its supremacy; but the implications of Hiroshima were now omnipresent. The very fact of Britain's emergence as a champion of the peaceful uses of nuclear energy was indissolubly linked to the technologies that had decimated Japan. However energetic the attempt to divorce the 'new nuclear age' from weapons of mass destruction, popular perceptions continued to be heavily influenced by 'natural' connections between 'atoms for peace' and 'atoms for war'. Yet among politicians, academics and research workers, scientistic and triumphalist dogma continued to legitimate facile distinctions between the aggressive and peaceful applications of nuclear energy. Convinced that the new form of energy would eventually prove itself economically and politically viable, zealous supporters of a nuclear future deployed the 'peaceful atom' as a means of optimistically counterbalancing the ever-present spectre of mass destruction. At no time, paradoxically, was this tendency more marked than during the early days of the Campaign for Nuclear Disarmament. Deeply committed to the rational application of scientific knowledge to human affairs, CND's leaders pointed to the peaceful deployment of the new source of energy both as a symbol of hope and as an exemplar of what *might* be achieved if nations could be persuaded to renounce the use of armed violence, 'deterrence' and blackmail.[40] This is not the place to describe the recent history of the anti-nuclear movement.[41] But it was only in the late 1960s that increased public awareness of the fallibility of nuclear technology gradually began to erode the political and psychological divide between 'peaceful' and 'aggressive' uses.[42]

Yet among unreconstructed triumphalists neither the safety issue nor the appalling economic record of nuclear electricity reduced commitment to the totalising social visions and 'power utopias' that have lain at the heart of this book. 'Unbelievers' could still be treated, as they had been in the 1920s, as fools or knaves. Even as intelligent and scientifically distinguished an advocate of nuclear power as Lord Marshall, the chairman of the Central Electricity Generating Board, would show himself capable of lecturing an audience in tones reminiscent of an editorial in *Electrical Industries and Investments*. In *Your Radioactive Garden*, Lord Marshall wrote that

> in order to guarantee adequate, safe and economical supplies the Central Electricity Generating Board must have transmission lines and power stations, including some nuclear stations. However, Mr or Mrs Public says: 'I want all the electricity I need but *no* pylons in my view, *no* power stations near my home – and above all, *no* nuclear waste in my back yard: put it somewhere else.' This NIMBY (Not In My Back Yard) syndrome applies to many other things as well – motorways and airports are examples – but its application to nuclear waste is very special because people *fear* nuclear waste. I think that fear is unjustified. People need to respect nuclear waste but not to fear it.[43]

Following a step-by-step, and graph-by-graph attempt to desensitise the needlessly anxious, Marshall concluded that

> the potential risks from . . . long-lived intermediate-level waste and high-level waste, put at a suitable depth in your garden, are similar to those from coal ash mixed in the top metre of your garden . . . all the other wastes I have discussed are considerably safer than that.[44]

The agency responsible for the CEGB's impressive 1980s television advertisement series – 'Energy for Life' – is more subtle and has perfected iconographies which have been central to triumphalist ideology and publicity since the early, heroic days of the Electrical Development Association. Camera movements, images, and rhythmic but conservative, orchestral rock music are deployed to underscore and amplify the cosmic and world-transforming status of electricity – and especially nuclear electricity. In a sample script, 'the sounds of the elements' give way to a bruised and menacing sky suddenly made resplendent by interweaving flashes of other-worldly lightning. The voice-over intones, in classically triumphalist style, that 'it cannot easily be explained. It cannot be seen or heard. It cannot be touched and yet' – here there is a subliminal shot deep into the bowels of a reactor – 'it can harness nature'. Then the camera tracks, ducking and weaving, over water and a deliberately mysterious landscape. This is a source of energy, the commentary continues, that can 'transform darkness at the speed of light'. Now we are in a skyscrapered and futuristic urban environment. A hygienically suited engineer-administrator points to a wall-chart and computer screen and prepares the way for a four-shot finale: a motorway aerially photographed at night but imaged as a solid stream of curving light; the moon surrealistically enlarged, like a massive, veined tennis ball, peeping out from behind an inner-city office block; three phallic buildings – power stations, government ministries, technocratic command centres? – and, finally, those same buildings hazily and magically transformed into a massive three-point plug. The commentary pushes home a message uncannily reminiscent of inter-war rhetoric: 'A most efficient and versatile power we can command as our *servant*. It is electricity. ENERGY FOR LIFE.'[45]

In another sixty-second sequence triumphalism is integrated with the familiar themes of domesticity, social stability and 'Englishness'. Near to dawn, a village, protected by hill and woodland, is at rest. In a bedroom, in half-light, a couple are sleeping. 'Every night while you're asleep a miraculous power is at work in the land ... a power which is used for everything from printing your morning paper to baking your daily bread ... ' Images reinforce these socially vital, though 'silent' and 'invisible' processes: this is a 'power so versatile that it can milk cows ... light our cities ... even sort the morning mail'. As the functions of electricity are rhythmically listed, the camera cuts back to the sleeping couple. Night finally gives way to day and the concept of 'service' is again introduced. 'Long before you wake, your electricity board is working for you. Drawing on its massive resources to serve you through the day.

Providing you with news, entertainment, light, heat and countless labour-saving devices.' A glowing, red computerised numeral on a television control panel, and an obliquely shot microwave oven reaffirm a long-established theme, still potent in the nuclear age: the rationalising and 'life-ordering' claims of electricity. The advertisement closes again, with the transformation of three buildings into a three-point plug.[46]

Half a century earlier, in his sound-track for Paul Rotha's documentary *The Way to the Sea*, W. H. Auden had ruminated on the same social, psychological and technological themes and meanings subliminally displayed in 'Energy for Life':[47]

> The lines wait
> The trains wait
> The drivers are waiting
> Waiting for Power.

Entering a classically 'thirties' urban landscape – 'A signal box. A power station' – Rotha's sea-bound locomotive speeds through 'the areas of greatest congestion; the homes of those who have the least power of choice'. Then, leaving the dirt and deprivation of the inner city, it approaches 'the first trees, the lawns and the fresh paint; district of the bypass and the season ticket'. Here, in the new suburbia:

> Power which helps us to escape is also helping those who cannot get away just now,
> Helping them to keep respectable,
> Helping them to impress the critical eye of a neighbour,
> Helping them to entertain their friends,
> Helping them to feed their husbands swept safely home each evening as the human tide recedes from London.

As the travellers move on into a social and cultural no man's land – that 'soulless', quasi-industrial grey area, feared and detested by Vaughan Cornish – Auden confronts the evidence and artefacts of the 'electrical revolution'.

> White factories stand rigid in the smokeless air.
> The pylon drives through the sootless field with power to create and to refashion.
> Power to transform on materials the most delicate and the most drastic operations.

Electrical technology, so visually prominent in an otherwise uninhabited and lifeless terrain between town and country, possesses the power to 'cleanse' and 'illuminate', 'lessen fatigue' and 'move deep cutters, milkers or separators'. Now that they have arrived at the sea, the day-trippers are urged to relax and discard every anxiety and inhibition: to become, quite literally, 'different people'. Be 'extravagant', Auden tells them, be 'lucky', 'clairvoyant' and 'amazing'. The travellers respond. Yet always at the end of the day, there must be the inevitable return to 'civilisation', inequality, and that most characteristically Audenesque of conditions, 'neurosis'. 'Power', electric or otherwise, has facilitated brief and intoxicating liberation but, despite everything the triumphalists and technological progressives may say or do, it does not, and cannot *in itself* transform the quality and texture of social relations, social hierarchies and everyday material life. That is a delusion which can plunge you down into an abstract and apolitical – ultimately a deeply and dangerously regressive – scientism. Mere 'mechanical' power is only one among many variables and problematics: it does not, in itself, constitute a 'solution'. In his ambivalent, pessimistic, and over-eclectic way, Auden understood that better than anyone.

Night falls and the tired day-trippers make their way slowly back to the train:

> The spectacle fades
> The tidy lives depart with their human loves.
> Only the stars, the oceans and the machines remain:
> The dark and the involuntary powers.

## Notes

1 'Campaign Launched to Cut Electricity Accident Toll', *Guardian*, 26 April 1988.
2 The cancer evidence has been well summarised in 'Electricity – a Shock in Store', *Panorama*, BBC TV, 21 March 1988 and the alleged connections between pylon 'fields' and neurological and psychological disorder assessed in 'Laying it on the Power Line', *Guardian*, 24 October 1984.
3 *A Programme of Nuclear Power* (HMSO, 1955), 1. For background to British nuclear power and policy-making in the period discussed in this chapter see Margaret Gowing, *Independence and Deterrence: Britain and Atomic Energy 1945–52* (1974); Duncan Burn, *Nuclear Power and the Energy Crisis* (1978); Roger Williams, *The Nuclear Power Decisions (1980)*; Walter C. Patterson, *Going Critical: an Unofficial History of British Nuclear Power* (1985) and the same author's *Nuclear Power* (revised edition, 1986); Tony Hall, *Nuclear Politics: the History of Nuclear Power in Britain* (1986); and Ian Welsh, 'British Nuclear Power: Legitimation and Protest 1945–1982' (University of Lancaster, PhD, 1988).
4 *A Programme of Nuclear Power*, 1.
5 Ibid. See, also, page 5 of the same document.
6 Ibid, 9.

7   Ibid, 11.
8   Ibid, 9.
9   Ibid, 11.
10  Ibid, 12.
11  *Hansard*, 25 February 1955, 1628. See, also, his comments in col. 1629.
12  Ibid, 1633.
13  Ibid, 1634.
14  Ibid, 1646.
15  Ibid, 1666.
16  Ibid, 1669–77.
17  PRO POWE 14/869. Brian Harrison to Duncan Sandys. 5 January 1956.
18  POWE 14/869. Aubrey Jones to Duncan Sandys. 27 January 1956.
19  POWE 14/869. 'Brief for the Minister's Meeting with a Deputation from Bradwell-on-Sea: Tuesday 28th February 1956.'
20  POWE 14/869. 'Record of a Deputation received by the Minister for Prime Minister on Tuesday, 28th February, about the Siting of an Atomic Power Station at Bradwell-on-Sea.'
21  POWE 14/869. N. J. Gibson to Clerk of the Parish Council, Bradwell. 6 March 1956.
22  POWE 14/869. Brian Harrison to Duncan Sandys. 5 January 1956.
23  POWE 14/869. Aubrey Jones to Brian Harrison. 12 March 1956.
24  Central Electricity Generating Board. Bankside House Archive. *Minutes of Proceedings at a Public Inquiry into an Application by the Central Electricity Authority to Establish a Nuclear Power Station at Bradwell-on-Sea*, 55.
25  Ibid, 62.
26  Ibid, 64.
27  Ibid, 30–42.
28  *Minutes* (Hearing of 8 May 1956), 7.
29  Ibid, 8.
30  Ibid, 11.
31  Ibid, 17. Driberg's words were quoted back at him by counsel for the Central Electricity Authority.
32  Ibid, 69. Evidence of M. V. Osmond.
33  *Minutes* (Hearing of 9 May 1956), 27.
34  Ibid, 30.
35  Ibid, 31.
36  Ibid, 32.
37  Ibid, 34.
38  POWE 14/869. D. E. H. Peirson to A. C. Campbell. 8 May 1956.
39  POWE 14/869. Duplicate of Cabinet. Home Affairs Committee memorandum, 'Siting at Bradwell-on-Sea of the First Nuclear Power Station in the White Paper Programme. Memorandum by the Minister of Fuel and Power'. n.d.
40  But, by the 1970s, CND was belatedly alerting itself to the dangers of nuclear power. See Paul Byrne. *The Campaign for Nuclear Disarmament* (1988), 122–3.
41  The background is sketched in by Tony Hall, *Nuclear Politics*, 131–47.
42  This process is described in Walter C. Patterson, *Nuclear Power*, Part 2.

43 Lord Marshall of Goring, *Your Radioactive Garden* (CEGB, 1986), 2.
44 Ibid, 16.
45 'Genesis'. CEGB Corporate Advertising Campaign (1986).
46 'Morning'. CEGB Corporate Advertising Campaign (1986).
47 Donald Mitchell, *Britten and Auden in the Thirties*, 90–3.

# A note on sources

Every chapter of this book – and particularly the first, on the ideological formation and texture of electrical triumphalism – has been based on a close reading of the technical press: the *Electrical Times, Electrical Review, The Electrician,* and least compromising of all in its attitude towards 'humanists' and 'doubters', *Electrical Industries and Investments*. This has been augmented by the official organ of the Electrical Association for Women, the *Electrical Age for Women* (later the *Electrical Age*) and Richard Borlase Matthews's crusading *Electro-Farming* (later *Electro-Farming and Rural Electrification*). The numerous publications of the Electrical Development Association – handbills, booklets, pamphlets, posters and reprints – currently preserved at the Electricity Council have also been drawn upon to convey the full flavour of electrical progressivism. One can only hope that the Council's archival activities will not be brought to a brutal halt following privatisation.

The chapter on the official imagery of electricity during these years rests on an analysis of EDA publicity, 'propaganda' and films. The latter are in excellent condition and can be viewed, on request, in cassette form at the Council. The internal workings of the Association are hinted at in successive *Reports to the Annual General Meeting* and more fully covered in its *Minutes*. Day-to-day relations with the Central Electricity Board are less clearly defined than they might have been, but fragments of information can be gleaned from the generally over-bland CEB *Minutes*, also held at the Electricity Council.

Caroline Haslett's life, work and lengthy directorship of the Electrical Association for Women are well documented at the Institution of Electrical Engineers. (See, in particular, NAEST 93/6.) EAW books, pamphlets and publicity hand-outs are also available in this collection, and, as an added bonus, the IEE Library boasts rich holdings of books and journals relating to the classic period of electrical progressivism and popularisation during the 1920s and the 1930s. (The library also provides courteous and unstuffy service to every type of reader.) Like all such sources, the annual *Reports* of the EAW are less enlightening than one feels they ought to be.

The scholarly study of patterns of urban supply during this period is still in its infancy and will certainly not reach adolescence if what happened in Manchester – the administrative papers of the Electricity Department as well as the *Minutes* of the Electricity Committee were discarded after nationalisation – is found to have been duplicated in other parts of the country. Since the cut-and-thrust of debates on electricity in the inter-war years was comprehensively censored in the formal *Proceedings* of the City Council, the historian is forced to rely on reports, and editorial comment, in the daily and weekly press. But this was the golden age of the provincial newspaper – Manchester carried no fewer than four reliable monthly reports of what went on in the Council chamber. If some of this journalism was less incisive than it might have been, one publication at least, the *Manchester and Salford Woman Citizen*, the organ of the local Women Citizens' Association, kept a weather eye on all matters electrical and environmental. Reading between the lines, one can also gain valuable information from the Manchester Corporation's annual handbook, *How Manchester Is Managed*. The Local History section of the Central Reference Library holds a small amount of valuable Electricity Department publicity material from the later 1930s.

Precisely because it was viewed by progressives as the most backward of electrical sectors, rural electrification was obsessively written and argued about. Richard Borlase Matthews's *Electro-Farming and Rural Electrification* proved itself a particularly rich source and was augmented by articles and pamphlets by a highly articulate lobby, which included, among others, Margaret Partridge, S. E. Britton and F. E. Rowland. The EDA also published a large number of publications, focusing mainly on 'progressive' areas such as 'rationalised' poultry and dairy farming, storage, canning and marketing.

The portrait of the anti-pylon movement was derived from local newspapers, central government records and the massive Archive of the Council for the Preservation of Rural England, which is held at the Museum of English Rural Life at the University of Reading. The *Sussex Express, East Sussex News* and *Sussex Daily News* carried large quantities of information on the Downland saga. The government's – and particularly Herbert Morrison's – strategy was reconstructed from material in PRO CAB 23 and 24, POWE 12/258 and the Ramsay MacDonald Papers (PRO 30/69/564). As for the account of ideology and action in the preservationist camp, this was based on CPRE 109/5/1,2 and 3 and CPRE 237. The Lakeland conflict proved to be more elusive – a visit to Keswick on a flawless day in early summer yielded less in terms of research than a sense of general well-being – but the Cumbrian Record Office at Kendal holds some of the records of the Society for Saving the Natural Beauty of the Lake District (CRO WDX/422/2/9) and the intermittently revealing *Minutes* of Keswick's confused and much-divided Council. All this was checked against accounts contained in the *Westmorland and Cumbrian Times*, *Cumberland and Westmorland Herald* and *West Cumberland Times*. As for the 'radicalisation' of Kenneth Spence, this is contained in CPRE 109/16.

The confrontation between the National Government (of 1931–5) and the New Forest protesters can be followed in the *Minutes and Correspondence of the Court of Verderers*, which are held in semi-catalogued form in the exemplary and, in architectural terms, beautifully appointed Hampshire County Record Office at Winchester. The most informative files proved to be HRO 7 M75 218X and 7 M75 267X. The Forestry Commission's somewhat Machiavellian role is documented in PRO F1/4 and PRO MAF 50/63, and governmental indecision in CAB 23 and 24. Local press coverage – in the *Bournemouth Daily Echo*, *Hampshire Advertiser and Southampton Times* and *Lymington and Milton Chronicle* – provided the regional context.

Baldwin and his ministers' attitudes towards – and Inskip's agonising over – the Caledonian Scheme are documented in CAB 23 and 24. (The Carbide Committee's proceedings are in CAB 16/174.) A large selection of the pamphlets, circulars and photographic evidence produced by the Association for the Preservation of Rural Scotland can be found in CPRE 102/8 and the views of Robert Boothby, Tom Johnston and other leading parliamentarians between the first bill in 1936 and the establishment of the North Scotland Hydro-Electric Board in 1943 are detailed in *Hansard*. This conflict was also played out at length in leaders and letters in *The Times*. For the Battersea story I was heavily dependent on Russell Moseley's research in PRO POWE 12/140, 12/141, 12/161 and 12/231. The King's intervention is fully minuted in POWE 12/140, and CAB 23 and 24 and the contemporaneous confrontation in *The Times* was followed up in the correspondence columns throughout April and early May 1929. Valuable background material and comment were also obtained from the technical press.

## Note on sources

There is little need to comment on the sources which provided the basis for Chapter 9, but the two volumes by Clough Williams Ellis – *England and the Octopus* (1928) and his edited collection, *Britain and the Beast* (1937) – are essential to any understanding of electricity and rural preservationism in the inter-war years. For the final chapter I made use of the crucial White Paper, *A Programme of Nuclear Power* (HMSO, 1955. Cmd 9389); the *Minutes* of the Bradwell public inquiry in 1956 held at the CEGB Archive at Bankside House, Southwark; and selected scripts from the CEGB 'Energy for Life' series, available on request from the Electricity Council. All the Council's historical archives have recently been transferred to the Greater Manchester Museum of Science and Industry.

# Select bibliography

Unless otherwise stated, place of publication is London.

Patrick Abercrombie and others, *The Coal Crisis and the Future: a Study of Social Disorders and their Treatment* (1926).

Patrick Abercrombie, *The Preservation of Rural England: the Control of Development by means of Rural Planning* (Liverpool, 1926).

Patrick Abercrombie, *Town and Country Planning* (1933).

Patrick Abercrombie, *Planning in Town and Country: Difficulties and Possibilities* (Liverpool and London, 1937).

Patrick Abercrombie, 'Country Planning' in Clough Williams Ellis (ed.), *Britain and the Beast* (1937), 133–40.

H. H. Ballin, *The Organisation of Electricity Supply in Great Britain* (1946).

John Barnicoat, *A Concise History of Posters* (1972).

Robert Boothby, *I Fight to Live* (1947).

Sue M. Bowden, 'The Consumer Durables Revolution in England 1932–1938: a Regional Analysis', *Explorations in Economic History*, 25, 1988, 42–59.

H. E. Bracey, *Industry and the Countryside: the Impact of Industry on Amenities in the Countryside* (1963).

Asa Briggs, *History of Broadcasting in the United Kindom: II: The Golden Age of Broadcasting* (1965).

Duncan Burn, *Nuclear Power and the Energy Crisis* (1978).

I. C. R. Byatt, *The British Electrical Industry 1875–1914: the Economic Returns of a New Industry* (Oxford, 1979).

Sir Norman Chester, *The Nationalisation of British Industry 1945–51* (1975).

Rob Cochrane, *Landmark of London: the Story of Battersea Power Station* (n.d., probably 1984).

Adam Collier, *The Crofting Problem* (1953).

T. A. B. Corley, *Domestic Electrical Appliances* (1966).

Vaughan Cornish, *National Parks and the Heritage of Scenery* (1930).
Vaughan Cornish, *The Poetic Impression of Natural Scenery* (1931).
Vaughan Cornish, *The Preservation of our Scenery: Essays and Addresses* (1937).
Ruth Schwartz Cowan, 'Two Washes in the Morning and a Bridge Party at Night: The American Housewife between the Wars', *Women's Studies*, 3, 1976, 147–72.
Ruth Schwartz Cowan, 'The "Industrial Revolution" in the Home: Household Technology and Social Change in the Twentieth Century', *Technology and Culture*, 17, 1976, 1–23.
Ruth Schwartz Cowan, *More Work for Mother: the Ironies of Household Technology from the Open Hearth to the Microwave* (New York, 1983).
Valentine Cunningham, *British Writers of the Thirties* (Oxford, 1988).
James Curran and Vincent Porter (eds), *British Cinema History* (1983).
Jennifer Dale, 'Class Struggle, Social Policy and State Structure: Central–Local Relations and Housing Policy' in Joseph Melling (ed.), *Housing, Social Policy and the State* (1980), 194–223.
Gerald Dix, 'Patrick Abercrombie 1879–1957' in Gordon E. Cherry (ed.), *Pioneers in British Planning* (1981), 103–30.
Mary Douglas, *Purity and Danger, an Analysis of Concepts of Pollution and Taboo* (1966).
Mary Douglas and Aaron Wildavsky, *Risk and Culture: an Essay on the Selection of Technological and Environmental Dangers* (Berkeley, Calif., 1983).
Elsie Elmitt Edwards, *Report on Electricity in Working Class Homes* (1935).
Clough Williams Ellis, *England and the Octopus* (1928).
Clough Williams Ellis, 'The Modern Landscape' in 'H.H.P.' and 'N.L.C.' (eds), *The Face of the Land* (1930), 11–15.
Clough Williams Ellis (ed.), *Britain and the Beast* (1937).
Richard Forty, *Objects of Desire: Design and Society 1750–1980* (1986).
E. W. Gilbert, 'Vaughan Cornish (1862–1948) and the Beauty of Scenery' in *British Pioneers in Geography* (Newton Abbot, 1972), 227–52.
Andrew Goudie, 'Vaughan Cornish – Geographer', *Transactions of the Institute of British Geographers*, 55, 1972, 1–16.
Jurgen Habermas, *Legitimation Crisis* (1976, translated by Thomas McCarthy).
Stuart Hall, 'John Grierson and the Documentary Film Movement' in James Curran and Vincent Porter (eds), *British Cinema History* (1983), 99–112.
Tony Hall, *Nuclear Politics: the History of Nuclear Power in Britain* (1986).

Leslie Hannah, *Electricity Before Nationalisation: a Study of the Development of the Electricity Supply Industry in Britain to 1948* (1979).
Leslie Hannah, *Engineers, Managers and Politicians: the First Fifteen Years of Nationalised Electricity Supply in Britain* (1982).
Christopher Harvie, *No Gods and Precious Few Heroes: Scotland 1914–1980* (1981).
Caroline Haslett, (ed.), *Electrical Handbook for Women* (1934).
Caroline Haslett, *Household Electricity* (1939).
Caroline Haslett, *Problems Have no Sex* (1949).
Delores Hayden, *The Grand Domestic Revolution: a History of Feminist Designs for American Homes, Neighborhoods and Cities* (Cambridge, Mass., 1981).
R. A. S. Hennessey, *The Electric Revolution* (1972).
Bevis Hillier, *Posters* (1969).
Thomas P. Hughes, *Networks of Power: Electrification in Western Society 1880–1930* (Baltimore, 1983).
Pauline Kael, *Kiss Kiss Bang Bang: Film Writings 1965–1967* (1970).
Pauline Kael, *5001 Nights at the Movies* (1982).
David S. Landes, *The Unbound Prometheus: Technological Change and Industrial Development in Western Europe from 1750 to the Present* (Cambridge, 1969).
Andrew Lees, *Cities Perceived: Urban Society in European and American Thought 1820–1940* (Manchester, 1985).
C. Day Lewis (ed.), *The Mind in Chains: Socialism and the Cultural Revolution* (1937).
David Lowenthal and Hugh C. Prince, 'The English Landscape', *Geographical Review*, 54, 1964, 325–31.
John Lowerson, 'Battles for the Countryside' in Frank Gloversmith (ed.), *Class, Culture and Social Change: a New View of the 1930s* (Brighton, 1980), 258–80.
Edward D. MacDonald (ed.), *Phoenix: the Posthumous Papers of D. H. Lawrence* (1936).
Roy and Kay MacLeod, 'The Social Relations of Science and Technology, 1914–1939' in Carlo M. Cipolla (ed.), *The Fontana Economic History of Europe: the Twentieth Century: Vol. I* (1976), 301–63.
Manchester Women's History Group, 'Ideology and Bricks and Mortar – Women's Housing in Manchester Between the Wars', *North-West Labour History*, 12, 1987, 24–48.
David Marquand, *Ramsay MacDonald* (1977).
Richard Borlase Matthews, *Electro-Farming or the Application of Electricity*

to *Agriculture* (1928).
J. Mayers and B. Spiers (eds), *Where Do We Go from Here?* (1938).
Rosalind Messenger, *The Doors of Opportunity: a Biography of Dame Caroline Haslett* (1967).
Keith Middlemas and John Barnes, *Baldwin: a Biography* (1969).
Donald Mitchell, *Britten and Auden in the Thirties: the Year 1936* (1981).
Herbert Morrison, 'The Elected Authority – Spur or Brake?' in Royal Institute of Public Administration, *Vitality in Administration* (1957), 7–17.
Charles Loch Mowat, *Britain between the Wars 1918–1940* (1955).
Anthony Passmore, *Verderers of the New Forest: a History of the Forest* (Old Woking, n.d.)
Walter C. Patterson, *Going Critical: an Unofficial History of British Nuclear Power* (1985).
Walter C. Patterson, *Nuclear Power* (revised edition, 1986).
Sidney Pollard, *The Development of the British Economy 1914–1980* (third edition, 1983).
Hugh Quigley, 'The Highlands of Scotland: Proposals for Development', *Agenda: A Quarterly Journal of Reconstruction*, 3, 1944, 77–96.
Wilfred L. Randell, *Electricity and Women: 21 Years of Progress* (1946).
Arthur Redford (assisted by I. S. Russell), *The History of Local Government in Manchester: III: The Last Half Century* (1940).
Thomas Regan, *Labour Members of the City Council 1894–1965* (Manchester, typescript, 1966).
Eric Rhode, *A History of the Cinema from its Origins to 1970* (1976).
W. G. Rimmer, 'Leeds and its Industrial Growth: Gas and Electricity (ii)', *Leeds Journal*, 28, 1957, 299–303.
W. A. Robson (ed.), *Public Ownership: Developments in Social Ownership and Control* (1937).
W. A. Robson (ed.), *The Political Quarterly in the Thirties* (1971).
J. Rodger, 'Inauthentic Politics and the Public Inquiry System', *Scottish Journal of Sociology*, iii, 1978, 103–27.
Mark H. Rose and John Clark, 'Light, Heat and Power: Energy Choices in Kansas City, Wichita and Denver, 1900–1935', *Journal of Urban History*, 5, 1979, 340–64.
F. R. Sandbach, 'The Early Campaign for a National Park in the Lake District', *Transactions of the Institute of British Geographers*, 3, 1978, 498–512.
Peggy Scott, *An Electrical Adventure* (n.d., probably 1934).
John Sheail, *Rural Conservation in Inter-War Britain* (Oxford, 1981).

Michael J. Shiel, *The Quiet Revolution: the Electrification of Rural Ireland 1946–1976* (Dublin, 1984).
E. D. Simon, *City Council from Within* (1926).
E. D. Simon and Marion Fitzgerald, *The Smokeless City* (1922).
Shena D. Simon, *A Century of City Government: Manchester 1838–1938* (1938).
Michael Stenton and Stephen Lees, *Who's Who of British Members of Parliament: Vols III and IV* (Brighton, 1979 and 1981).
John Stevenson, *British Society 1914–45* (1984).
Lawrence Stone and Jeanne C. Fawtier Stone, *An Open Elite? England 1540–1880* (Oxford, 1984).
R. Stone, *Measurement of Consumers' Expenditure in the United Kingdom 1920–1938* (Cambridge, 1954).
W. E. Swale, *Forerunners of the North Western Electricity Board* (Manchester, 1963).
A. J. P. Taylor, *English History 1914–45* (Oxford, 1965).
F. M. L. Thompson, *English Landed Society in the Nineteenth Century* (1963).
Edmund N. Todd, 'A Tale of Three Cities: Electrification and the Structure of Choice in the Ruhr, 1886–1900', *Social Studies of Science*, 17, 1987, 387–412.
G. M. Trevelyan, *Must England's Beauty Perish? A Plea on Behalf of the National Trust for Places of Historic Interest or Natural Beauty* (1929).
G. M. Trevelyan, *Clio: a Muse and Other Essays* (1929).
G. M. Trevelyan, 'Amenities and the State' in Clough Williams Ellis (ed.), *Britain and the Beast* (1937), 183–6.
G. M. Trevelyan, *English Social History: a Survey of Six Centuries: Chaucer to Queen Victoria* (1942).
G. M. Trevelyan, *An Autobiography and Other Essays* (1949).
David Turnock, *Patterns of Highland Development* (1970).
Martin J. Weiner, *English Culture and the Decline of the Industrial Spirit* (Cambridge, 1981).
Ian Welsh, 'British Nuclear Power: Legitimation and Protest 1945–1982' (University of Lancaster, PhD, 1988).
Gary Werskey, *The Visible College: a Collective Biography of British Scientists and Socialists of the 1930s* (1978).
Edith M. Whetham, *Agrarian History of England and Wales: VIII: 1914–1938* (Cambridge, 1978).
Hubert Williams (ed.), *Man and the Machine* (1935).
Raymond Williams, *The Country and the City* (1973).
Roger Williams, *The Nuclear Power Decisions* (1980).

Andrew Wilson, 'The Strategy of Sales Expansion in the British Electricity Supply Industry between the Wars' in Leslie Hannah (ed.), *Management Strategy and Business Development* (1976), 203–12.

R. E. Wraith and G. B. Lamb, *Public Inquiries as an Instrument of Government* (1971).

Brian Wynne, *Rationality and Ritual: the Windscale Inquiry and Nuclear Decisions in Britain* (British Society for the History of Science Monographs, 3, 1982).

# Index

Abercrombie, Sir Patrick, 5, 96, 164, 168
  anti-urbanism, 165
  disillusionment with purist preservationism, 167
  early idealisation of the English village, 165
  education and formative years, 157
  scepticism towards state environmental policies, 166–7
  theory of the 'natural' and the 'technological', 166
accents in British cinema, 38 n.62
antiurbanism as an ideology, 167
*Architect's Journal*, 101
Ashley, Wilfred, 139, 143
Association for the Preservation of Rural Scotland (APRS), 121, 127, 128
Atomic Energy Authority, 173, 178
Auden, W. H., 12, 182–3

Baldwin, Stanley, 111, 127
Battersea Borough Council, 141
Battersea Power Station, 4
  backlash against environmentalist critique, 148–9
  criticisms of Londonderry concessions, 145–6
  debate in House of Lords, 144–5
  early plans, 138
  George V's disquiet, 142–3
  investigations by Government Chemist, 151–2
  Office of Works seeks assurances, 140
  opposition of metropolitan boroughs, 140–1
  provisional solution of sulphur problem, 151–2
  public fear of sulphur fumes, 141
  Savoy Hill hearing, 139–40
Beauchamp, J. W., 23, 40
Bedford, 81
Belloc, Hilaire, 96
Betjeman, John, 174
Birmingham, 40
Bishop's Stortford, 157
Blomfield, Sir Reginald, 98, 113
Blunt, Reginald, 141, 142
Booth, Mrs H., 139
Boothby, Robert, 125–6, 130, 132
Bradwell inquiry into construction of atomic power station (1956)
  inspector's report, 178
  sanguine official attitude, 175–6
  Tom Driberg's critique, 176–7
Bradwell-on-Sea, 174, 175, 176
British Aluminium Company, 118
British Oxygen Company (BOC), 121, 122, 124, 125, 126, 127, 131
Britton, S. E., 80
Brown, W. A., 113
Butcher, John, 172
Buxton, Earl, 99

calcium carbide, 132, 135 n.25
Calcium Carbide Committee (1937), 126
Caledonian Scheme, first Bill (1936)
  arguments of opponents, 122
  arguments of supporters, 121
  defeat, 121; *see also* hydroelectricity, Scotland
Caledonian Scheme, second Bill (1937)
  arguments of opponents, 124–5
  governmental vacillation, 123–4

# Index

intervention by Robert Boothby, 125–6;
   *see also* hydroelectricity, Scotland
Caledonian Scheme, third Bill (1938)
   balance of opinion before debate, 128
   debate and defeat, 129–31
   opposition from environmental lobby,
      127–8; *see also* hydroelectricity,
      Scotland
Campaign for Nuclear Disarmament, 180
Cardiganshire, 76
Carmarthenshire, 76
Central Electricity Authority, 174
Central Electricity Board (CEB), 2, 19, 25,
   26, 95, 101, 102, 104, 105, 106, 107, 108,
   109, 111, 113, 114
Central Electricity Generating Board
   (CEGB), 191
Central Landowners' Association, 96
Chamberlain, Neville, 17, 143
Chandler, Montague, 111, 112, 113, 114
Chelsea Borough Council, 140–1, 150
Chelsea Society, 141
Cheshire, 80
Chesser, Elizabeth Sloan, 44
Chester, 80
Child, H. H., 108
Clyde Valley Electric Power Company,
   76
Cochran Boiler Company, 39
Cockermouth, 104
Coniston, 160
Cornish, Vaughan, 5, 166, 167, 168, 176,
   182
   attitude towards urbanism and
      suburbanism, 161–2
   education and early years, 156–7
   formulating a 'science of scenery', 159–60
   idealisation of village life, 161
   impact of English romanticism, 160–1
Corpach, 121, 124
Council for the Preservation of Rural
   England (CPRE), 95, 100, 102, 106, 107,
   108, 157, 163, 165, 177
Council for the Preservation of Rural
   Wales (CPRW), 163, 165
*Country Life*, 102
Court of Verderers, 109, 111, 113, 114
crafts and skills, rural, 84–6; *see also*
   revivalism, rural
Craig, Elizabeth, 44
Cramb, A. C., 15, 24
Crompton, Col R. E., 41
Curzon Howe, Lady, 112

*Daily Express*, 12, 67
*Daily Herald*, 98
dairy farming, 82
Davidson, John, 122, 125, 129
Davidson, Lord, 144
Dawson of Penn, Lord, 142, 151
De la Bere, Sir Rupert, 75
Department of Scientific and Industrial
   Research, 143, 152
Desmore, Violet, 44
Driberg, Tom, 174, 176; *see also* Bradwell
   inquiry (1956)
Duff, C. P., 98
Dumfriesshire, 76
Duncan, Sir Andrew, 25, 104, 132
Dykes, A. H., 108

East Sussex County Council, 95
education, electrical, 15–16, 172
   *see also* Electrical Association for
   Women; triumphalism, electrical
Edwards, Elsie Elmitt, 42, 53, 82
*Electrical Age* (subsequently *Electrical
   Age for Women*), 15, 44, 82
electrical appliances, 44–5
Electrical Association for Women (EAW),
   3, 35
   attitude towards working-class, 45
   early financial problems, 40
   foundation, 40
   reaction to the 'servant problem', 42,
      45–6
   relations with EDA, 40–1
   rumours of internal dissension, 40–2
   social origins of membership, 43; *see also*
      Electrical Development Association;
      electrical industry: publicity
Electrical Development Association
   (EDA), 3, 15, 19, 181
   organisation and funding, 23–4
   relations with CEB, 24–6; *see also*
      Central Electricity Board; electricity
      industry: publicity
*Electrical Industries and Investments*, 3, 74,
   79, 100
*Electrical Times*, 3, 23, 52, 148, 149
*Electrician*, 3, 13, 15, 64, 83, 84, 131, 148, 149
electricity
   aid to interior design, 43–4
   as 'social service', 12
   connections with rural preservationism,
      83–4
   rivalry with gas, 16–19

Electricity Commissioners, 2, 76, 80, 95, 138, 139, 140, 143
electricity industry: publicity
  cinematic techniques, 33–4
  health and cleanliness, 29
  interior design, 30
  middle-class bias, 27, 34–5
  rationalisation of kitchen routines, 30–1
  reducing domestic expenditure, 31–2
  refrigeration, 31–2
  replacement of servants, 27–9; *see also* Electrical Association for Women; Electrical Development Association; triumphalism, electrical
electricity supply, urban
  reasons for slow growth, 52–4
  spatial and regional distribution, 69 n. 13
electrification, rural
  in Eire, 77
  in Scotland, 76
  in Ulster, 77
  in Wales, 76
  international comparisons, 74
  relative retardation in Britain, 74–5
electro-culture, 73, 77–8; *see also* electrification, rural; triumphalism, electrical
Elliott, Walter, 123, 124, 126
Ellis, Clough Williams, 5, 165, 167, 168, 176, 178
  agenda for rural preservationism, 163–4
  education and early years, 157
  impact of *England and the Octopus*, 162–3

Fennell, W., 83
Fife Power Company, 76
Fladgate, W. F., 148–9; *see also* Battersea Power Station
Fletcher, T. P., 175–6
Fordingbridge, 111
Forestry Commissioners, 112, 113, 114, 115
Forster, E. M., 164
Forster, Lord, 111
Fort Augustus, 124
Fort William, 118, 121, 128
Foyers, 118

Gage, Lord, 96
Galsworthy, John, 96
Garry, River, 121
gas industry
  competition with electricity, 16–19
  role of 'visible flame' in effective cooking, 37 n. 46
General Electric Company, 12
George V, King, 142
Gill, Eric, 98
Glasgow, 40
Glen Garry, 128
Glenmoriston, 124
Griffin, H. G., 106, 107, 108, 109
Gwynne, Col R. V., 96

Hammond, John and Barbara, 159
Hannah, Leslie, 133
Hardie, George, 119
Harrison, Brian, 174
Haslett, Caroline, 3, 172
  awarded CBE, 41
  education and early career, 39–40
  writings on science and society, 47–9; *see also* Electrical Association for Women; triumphalism, electrical
Haward, Sir Harry, 140
Highland Development League, 128
high-tension wires
  alleged cause of cancer and depression, 172
Hill, Rev James, 124
Hines, G. C., 165
hydroelectricity, Scotland
  anti-English opposition to private schemes, 119
  early attempts at development, 118
  failure of public control, 118–9
  long-term problems, 133–4
  short-term social benefits, 133; *see also* Caledonian Schemes (1936), (1937), (1938)
'I'm Electric', 26
Inskip, Sir Thomas, 123, 127, 130–1
Institution of Electrical Engineers, 3, 23
*Interim Report of the Water Power Committee* (1919) (Snell Committee), 118–19; *see also* hydroelectricity, Scotland
Inverness, 121, 124

Jackson, W. T., 58
Joad, C. E. M., 164
Johnston, Tom, 132, 134
Jones, Aubrey, 175

Kensington, 41
Kensington Borough Council, 141, 150
Keswick Anti-Pylon Committee, 104,

## Index

105, 108, 109
Keswick pylon dispute (1929–33)
  'anti-electric' victory, 108
  campaign renewed, 107
  coordinated local opposition, 104
  reasons for defeat of CEB, 109; *see also* Central Electricity Board; Council for the Preservation of Rural England; Sherrard, O. A.
Keynes, J. M., 96, 164
Kinochleven, 118
Kipling, Rudyard, 96
Kirkwood, David, 119, 133

Laboratory of the Government Chemist, 143
Lamb, H. C., 58, 60
Leigh, 53
Llewellin, Col John, 132
Lloyd George, David, 74
London County Council, 150
Londonderry, Lord, 144–5; *see also* Battersea Power Station
*London Evening News*, 100
London Power Company (LPC), 138, 139, 140, 144, 145, 146, 149, 150, 151, 152
Lovell, Percy, 146

MacDonald, John Ramsay, 99, 113
Macdonald, Sir Murdoch, 122, 129
MacLeod of MacLeod, Flora, 129
MacMillan, Malcolm, 132
Manchester
  comparisons with other urban areas, 63, Table 3
  domestic electricity supply, 54–7, Table 2
  Labour Party and extension of supply, 58–60
  plight of working-class consumers, 58, 62
Manchester Corporation Electricity Department
  conflict over pricing and general rate, 60–1
  early history, 54
  sales ethos and strategies, 64–6
  moderate municipalism, 66–7
Manchester Corporation Gas Department
  conflict with Electricity Department, 68
  early developments, 68
  revived commercial success, 68–9
Marshall of Goring, Lord, 180

Matthews, Richard Borlase, 73, 74, 75, 77, 78, 79, 83, 84, 85; *see also* electro-culture
mechanisation and its opponents, 1
Mills, Major J. D., 111
Ministry of Labour, 81
Ministry of Transport, 2, 3, 80, 95
Moir, Lady, 43
Montagu, Lady, 112
Moriston, River, 121
Morrison, Herbert, 74, 96, 99–100, 145, 147; *see also* South Downs public inquiry (1929)
Mount Temple, Lady, 43
'Mr Therm', 24, 26
Municipal Reform Party, 25

National Trust, 105, 106, 107, 163
nature mysticism, 167
New Forest anti-pylon movement
  governmental indecision, 113
  origins of dispute, 111–12
  public inquiry into CEB proposals, 111–12
  reasons for success of protest, 114
New Forest Association, 109, 111–12
Newton, Sir Douglas (later Lord Eltisley), 75, 83
North Scotland Hydro-Electric Board, 132–3
Norwich, 80–1

*Observer*, 86
Office of Works, 140

Page, Sir Archibald, 25
Palmer, Arthur, 174
Partridge, Margaret, 52, 79
Paterson, Mrs C. C., 43
Peacehaven, 162
Pearce, S. L., 58, 139, 151
Peirson, D. E. H., 178
Penrith, 104
Portmeirion, 157
Port of London Authority, 144
Port Talbot, 127, 128, 131
poultry industry, 81–2
Pybus, P. J., 113–14, 115
pylons, public attitudes towards, 100–1

Quigley, Hugh, 60

Radnorshire, 76

Railway Commissioners, 2
Ramsay, Archibald, 121, 123
Ray, Sir William, 25, 78
Redmayne, Sir Richard, 146
Reepham, 81
Regan, Tom, 59, 60, 61, 62, 70 n.28
*Report on Electricity in Working Class Homes* (1935), 42
revivalism, rural, 83, 85
   failure in inter-war Britain, 86–7
Robinson, J. E., 61
Robinson, Sir Roy, 113
Robinson, W. A., 165
Rotha, Paul, 182
Rowland, F. E., 81
Runciman, Walter, 123–4
*Rural Electrification*, 85
Rural Industries Bureau, 84
rural patriciate
   collapse in inter-war Britain, 168, 170–1 n.77
Rural Reconstruction Association, 84
Ruskin, John, 160

Salford, 40
Scott, Sir Leslie, 111, 113, 114
Scottish Highlands, 4
Seager, J. E., 95
Shaw, Bernard, 58
Sherrard, O. A., 106, 107, 109, 114
Simon, Sir John, 127
Sloan, R. P., 24
Snell, Sir John, 139
Society for Saving the Natural Beauty of the Lake District (SSNBLD), 102
South Downs, 4
South Downs public inquiry (1929)
   advocates of electrification, 98
   reasons for defeat of protest, 98
   opposition to pylons, 96–8
   petition against CEB scheme, 98
   reasons for defeat of protest, 98, 101
Spence, Kenneth, 102, 105–6, 107, 108
Spender, Stephen, 12

state, local opposition to, 6
Stowe School, 157
sunlight, artificial, 18
Sussex Archaeological Society, 96
Sussex Downsmen, 95
Sutton, Cecil, 111–2, 114

*Times, The*, 98, 112, 127, 128–9, 131, 142, 146
Tizard, Henry, 152
Trevelyan, G. M., 5, 161, 164, 167, 168
   early years as historian, 156
   commitment to preservationism, 159
   cultural pessimism, 159
   love of walking and 'wilderness', 158–9
triumphalism, electrical
   animosity towards popular press, 14
   contrasted with 'millennarianism', 20 n.2
   definitions of, 10
   discourses associated with, 19–20
   links with 'futurology', 11
   relationship to patriotism, 12
   relevant pressure groups, 2–3; *see also* electricity industry: publicity
triumphalism, nuclear, 174–5
   converting 'unbelievers', 180
   publicity techniques, 181–2
Troubridge, Sir Thomas, 112

University College, London, 157
University of Liverpool, 157

Walker, William, 59, 60, 61, 70 n.26
Wallington, 157
Walpole, Hugh, 104
wayleave policies, 79–80
Wedderburn, H. J. S., 132
Weiner, Martin, 167
Welwyn Garden City, 52
Westminster, Duchess of, 112
Whinlatter Pass, 102
Willey, Fred, 174
Women's Engineering Society, 3, 40